Lectures in Elementary Probability Theory and Stochastic Processes

Lectures in Elementary Probability Theory and Stochastic Processes

Jean-Claude Falmagne

University of California, Irvine

Boston   Burr Ridge, IL   Dubuque, IA   Madison, WI   New York
San Francisco   St. Louis   Bangkok   Bogotá   Caracas   Kuala Lumpur
Lisbon   London   Madrid   Mexico City   Milan   Montreal   New Delhi
Santiago   Seoul   Singapore   Sydney   Taipei   Toronto

McGraw-Hill Higher Education

A Division of The McGraw·Hill Companies

LECTURES IN ELEMENTARY PROBABILITY THEORY AND STOCHASTIC PROCESSES

Published by McGraw-Hill, a business unit of The McGraw-Hill Companies, Inc., 1221 Avenue of the Americas, New York, NY 10020. Copyright © 2003 by The McGraw-Hill Companies, Inc. All rights reserved. No part of this publication may be reproduced or distributed in any form or by any means, or stored in a database or retrieval system, without the prior written consent of The McGraw-Hill Companies, Inc., including, but not limited to, in any network or other electronic storage or transmission, or broadcast for distance learning.

Some ancillaries, including electronic and print components, may not be available to customers outside the United States.

This book is printed on acid-free paper.

1 2 3 4 5 6 7 8 9 0 DOC/DOC 0 9 8 7 6 5 4 3 2

ISBN 0–07–244890–3

Publisher: *William K. Barter*
Senior sponsoring editor: *David Dietz*
Developmental editor: *Peter Galuardi*
Executive marketing manager: *Marianne C. P. Rutter*
Senior marketing manager: *Curtis D. Reynolds*
Senior project manager: *Jayne Klein*
Production supervisor: *Sherry L. Kane*
Coordinator of freelance design: *Rick D. Noel*
Cover designer: *Rokusek Design*
Cover image: *John Rokusek/Rokusek Design*
Supplement producer: *Brenda A. Ernzen*
Typeface: *11.5/14 Century*
Printer: *R. R. Donnelley & Sons Company/Crawfordsville, IN*

Library of Congress Cataloging-in-Publication Data

Falmagne, Jean-Claude.
  Lectures in elementary probability theory and stochastic processes / Jean-Claude
Falmagne. — 1st ed.
     p.   cm.
  Includes index.
  ISBN 0–07–244890–3
  1. Probabilities. 2. Stochastic processes. I. Title.

QA273 .F33 2003
519.2—dc21
                                    2002066017
                                    CIP

Some of the mathematicians who made essential contributions to the foundation of probability theory have their pictures on the cover of this book. Starting on the upper left, and going clockwise, we find Johann Bernoulli (1667-1748), Blaise Pascal (1623-1662), Jakob Bernoulli (1654-1705), and below him Siméon-Denis Poisson (1781-1840). The table at the bottom of the cover (and also on the back cover) is the so-called 'Pascal's triangle.' It is reproduced from the original edition of *"Traité du triangle arithmétique, avec quelques autres traités sur le même sujet,"* by Pascal, posthumously published in 1665 by Guillaume Deprez. In modern texts, this table is usually rotated 45 degrees to the right, as shown by the partial table below. The letters in the original table make sense only in the context of Pascal's text and are omitted. For the meaning of the table, see Equation (7.6) and Problem 11 of Chapter 7.

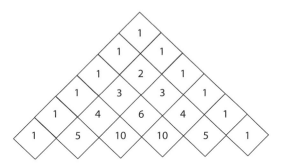

Contents

Preface

This monograph is largely based on a course on elementary probability given for a number of years to graduate students in the cognitive and other behavioral sciences at NYU and at UCI. With the first couple of weeks being devoted to a review of some basic mathematical facts, the first 18 chapters can be covered in a quarter course, and have been written so that each chapter corresponds to one fifty-minute lecture. Such a course is intended as a preparation for a more or less standard statistics sequence and for courses in mathematical modeling. The other chapters can be used as one sees fit for a semester course or as complements. Probability theory is covered here in considerably more detail than is usually the case for such students. Typically, this material is dealt with in a rather cursory manner, as if one should get as quickly as possible to what is regarded as the core subject, that is, statistics. The author finds such a practice objectionable because neither statistics, nor mathematical modeling, can go very far without a solid foundation in probability. Slighting it leads to a limited understanding of the subject which could promote superstitious scientific behavior.

Another unusual feature lies in the content of Chapter 1. We give there a collection of well-known examples of faulty probabilistic reasoning. It is a well documented fact that correct probabilistic reasoning does not automatically evolve as part of a sophisticated education if a special effort is not made to foster it. Scientifically educated people are capable of gross probabilistic reasoning blunders. Being aware of the most prominent fallacies may protect the students against some pitfalls and convince them that a serious study of the subject is worth the effort.

I am indebted to Lin Nutile for helping with some of the typing and the organization, and especially to Chris Doble for his detailed remarks on the manuscript and his work on the problems at the end of each chapter. Geoff Iverson read part of a draft and made, as always, useful comments, for which I thank him. I am also obliged to the graduate students who took a course based on this book in the fall of 2001 at UCI. They picked up many typos and errors. In the last stretch, Dina and Sophia F. accepted goodnaturedly the professional component of our summer vacation, and I am grateful to them for their always kind tolerance of my working schedule.

Jean-Claude Falmagne
Irvine, California

Chapter 1

Preliminaries

In some respects, the human mind is a rather crippled machine which, left to its own device, is liable to pitfalls of various sorts. This applies especially to probabilistic reasoning. We begin this monograph with a few examples of well documented reasoning fallacies. These examples indicate that these fallacies are not restricted to contrived experimental situations, and may affect decisions which may be crucial in people's lives, such as whether or not to administer a vaccine, to perform amniocentesis or to prepare for a war. They should motivate the reader for a serious study of our subject.

The Framing of Decisions

One prominent reason why rational decisions are not always attainable is that, when presented with a situation requiring a choice, we are often unduly influenced by the framework, in other words, by accidental features of the situation which should have no bearing on the decision. This may result in a failure to recognize as essentially identical two situations described in different terms. In the example below, probabilistic reasoning is required but is not central to the fallacy.

1.1. Example. (SURVIVAL FRAME VERSUS MORTALITY FRAME.) This example comes from a study by McNeil et al. (1982) of preferences between medical treatments. The subjects were provided with some statistical information concerning the outcomes of two modes of treating lung cancer: surgery and radiation therapy. To some respondents, the information was given in terms of survival rates, and

1

to the others, in terms of mortality rates. The subjects were then asked to make a hypothetical choice between the two treatments. The exact descriptions given to the subjects were as follows.

SURVIVAL FRAME

Surgery: Of 100 people having surgery, 90 live through the postoperative period, 68 are alive at the end of the first year, and 34 are alive at the end of five years.

Radiation Therapy: Of 100 people having radiation therapy, all live through the treatment, 77 are alive at the end of one year, and 22 are alive at the end of five years.

MORTALITY FRAME

Surgery: Of 100 people having surgery, 10 die during surgery or through the postoperative period, 32 die at the end of the first year, and 66 die by the end of five years.

Radiation Therapy: Of 100 people having radiation therapy, none die during treatment, 23 die by the end of one year, and 78 die by the end of five years.

A little reflection shows that the differences between the corresponding descriptions are superficial. Nevertheless, 18% (of 247 subjects) favored radiation therapy when the information was given in terms of survival rates. This percentage rose to 44% (of 336 subjects) when mortality rates were used. Remarkably, the framing effect held whether the subjects were clinical patients, experienced physicians, or even statistically sophisticated students from a business school.

This effect is by no means unique, nor even especially difficult to elicit. Sometimes, as shown by Tversky and Kahneman (1981), the results can be quite spectacular. Another kind of error arises in the so-called *Conjunction Fallacy* in which probabilistic reasoning is directly involved.

The Conjunction Fallacy

In everyday life, we often have to make decisions in situations in which the outcomes of our choices are uncertain. One of the most basic laws of probability theory is that the probability of the joint realization of two events, say A and B, cannot be larger that the probability of either of the two events, considered on its own. For example, the probability of having a blond child with blue eyes cannot be larger than the probability of having a blond child. Sometimes, these two probabilities may be equal.[1] A standard convention is to write

$$A \cap B$$

for the joint event that A and B are simultaneously realized in an experiment. Denoting the probabilities of the events by $\mathbb{P}(A)$, $\mathbb{P}(B)$ and $\mathbb{P}(A \cap B)$, the basic probability law under discussion thus takes the form

$$\mathbb{P}(A \cap B) \leq \mathbb{P}(A) \tag{1.1}$$

and similarly, of course, $\mathbb{P}(A \cap B) \leq \mathbb{P}(B)$. This law is sometimes called the *conjunction rule*. There are very strong intuitive reasons why the conjunction rule should be satisfied: Each time $A \cap B$ is realized in an experiment, A is also automatically realized (since both A and B are realized). Thus, A is realized at least as often as $A \cap B$, and the same argument applies to the event B.

One of the most pervasive cognitive fallacies consists in the violation of Eq. (1.1) in the judgment of conjunction probabilities: In some circumstances, people exhibit a strong tendency to judge a conjunction as more likely than one of its constituents. This may happen even with sophisticated individuals and in situations where the concrete details should a priori eliminate any possibility of a misunderstanding. Some researchers have suggested that subjects in this type of experiment may have a different interpretation of the concept of 'probability' or 'likelihood' than those used by professional statisticians. The experiment reported below was designed by Tversky and Kahneman (1983) to test that hypothesis. As we shall see, the data reject it.

1.2. Example. (THROWING A DIE.) The instructions to the subject, which are reproduced below, do not contain the words 'probability' or 'likelihood.'

INSTRUCTIONS. Consider a regular six-sided die with four red faces and two green faces. The die will be rolled 20 times and the sequence of greens (G) and reds (R) will be recorded. You are asked to select one sequence, from a set of three, and you will win \$25 if the sequence you choose appears on successive rolls of the die. Please choose the sequence of greens and reds on which you prefer to bet.

    (1) RGRRR   (2) GRGRRR   (3) GRRRRR

---

[1] For an eskimo woman who has always been faithful to her eskimo husband, both of these probabilities may be equal to zero.

Note that Sequence 1 can be generated by deleting the first G in Sequence 2. Thus, a subject choosing Sequence 1 would win if Sequence 1 or Sequence 2 is realized. Two hundred sixty students from Stanford and UBC were given this problem, 150 of them with real payoffs, and the rest with hypothetical payoffs. In both cases, a majority of the subjects choose Sequence 2. The percentages were 65% and 62%, respectively. The correct choice is of course Sequence 1. The error is an instance of the conjunction fallacy. The reader is invited to find out why this is the case (see Problem 1).

A plausible interpretation of such results, which seems to fit many data, is that the subjects have a natural tendency to gauge the three alternatives with respect to a *prototype*: Many respondents choose Sequence 2 because it is prototypical of the die used in the experiment: the ratio of the red and green in Sequence 2 is the same as that of the faces of the die.

In some situations, it is appropriate to talk about *scenarios* instead of prototypes. This applies to the following example.

1.3. Example. (SCENARIOS AND FORECASTS.) In still another experiment of Tversky and Kahneman (1983), the subjects were 115 participants in the Second International Congress on Forecasting, held in Istanbul, Turkey, in July 1982. They were, for the most part, professional analysts, working in industry, universities, or research institutes. Professional forecasting and planning was part of their work, sometimes using scenarios. The design of the experiment was the same as the one just described. One group of forecasters was asked to rate the probability of the event

> a complete suspension of diplomatic relations between the USA and the Soviet Union, sometime in 1983

whereas the other other group had to rate the probability of

> a Russian invasion of Poland, and a complete suspension of diplomatic relations between the USA and the Soviet Union, sometime in 1983.

Thus, the second event is a conjunction of the first with "a Russian invasion of Poland." Nevertheless the ratings of the probabilities were significantly lower for the first event than for the conjunction. The geometric means were .14% and .47%, respectively. These results are especially remarkable because the subjects were presumably sophisticated regarding probability judgments.

These examples should serve as a warning to students of probability and statistics. Being sophisticated in these matters does not automatically immunize individuals against the fallacies. In particular, the prototypical or scenario modes of thinking are very enticing, and one should guard against them as much as possible.

Fighting the Prototypical or Scenario Mode

Fortunately, there are ways of reasoning about probability statements which render the conjunction fallacy less likely or less prominent. In the case of the forecasting situation of Example 1.3, one could (painstakingly) imagine a number of scenarios which would result in of the breakdown of diplomatic relations between the USA and the former Soviet Union. The invasion of Poland by the Soviet Uniion would be only one of them. Conceivably, this analysis of the possible cases might reduce the impact of the conjunction fallacy.

In some cases, the conditions of the experiment naturally suggest an analysis of the possible cases. Our final example illustrates this point.

1.4. Example. (THE HEALTH SURVEY) We are still drawing from the same article by Tversky and Kahneman (1983). The subjects were given the problem displayed below. The average estimates are given in parentheses.

ABSTRACT SAMPLE

A health survey was conducted in a sample of adult males in British Columbia, of all ages and occupations. Please give your best estimates of the following values:

- What percentage of the men surveyed have had one or more heart attacks? (18%)

- What percentage of the men surveyed both are over 55 years old and have had one or more heart attacks? (30%)

The respondents were 147 statistically naive students at UBC. A detailed analysis indicated that 65% of them committed the conjunction fallacy, estimating the percentage for the conjunction strictly higher than that of the component "have had one or more heart attack." This offers nothing new. The conjunction fallacy has its usual strong hold.

However, another sample of 117 subjects in the same pool was given the following closely related problem:

SAMPLE OF 100 MALES

A health survey was conducted in a sample of 100 adult males in British Columbia, of all ages and occupations. Please give your best estimates of the following values:

- How many of the participants have had one or more heart attacks?

- How many of the participants both are over 55 years old and have had one or more heart attacks?

In this version, only (!) 25% of the 117 subjects tested committed the conjunction fallacy. This effect was replicated by the same authors with several other problems of the same type. Moreover, in the second problem, the proportion of violations of the conjunction rule was further reduced to 11% with another group of 360 subjects, who were asked to estimate the number of participants over 55 years of age, prior to their estimating the conjunctive event.

The Need for Crutches or Reasoning Tools

By this time, the reader is probably convinced that the fallacies exemplified in this chapter are not mere curiosities arising in the laboratories of social scientists, but illustrate critical failures of human reasoning, which may have a detrimental impact on people's lives. Many other examples could be given. Often, these examples involve probabilistic reasoning.

However, being aware of the potential fallacies and being competent in probability and statistics does not guarantee immunity. You may remember a story that made the headlines a while ago, concerning a game show. The participants had to choose one of three closed doors, labeled A, B, and C. They were told by the show host that behind two of the doors was a goat, while the third door was hiding an expensive car. The goal was thus to choose the door hiding the car. There was a catch, however. After a contestant had made his or her choice of a door, the host would open one of the two remaining doors, and reveal a goat. The contestant would then be given the option to switch. In other words, the contestant chooses door A. Next, the host opens door B, revealing a goat. Should the contestant switch to door C? At some point, the creator of the show published her solution to this problem: She recommended the switch. This created somewhat of an outcry. Apparently, she received hundreds of letters criticizing the proposed solution, with quite a number of them coming from professional mathematicians. But her solution was correct. It is actually not very difficult to prove that the probability of finding the car behind door C is higher than finding it behind the chosen door A, given the open door B revealing a goat. Why were all these mathematicians wrong? They certainly had all the tools needed to get the solution. A good guess is that they were careless and relied on their intuition, which in this case strongly suggests that the probabilities of finding the car behind door A or door C should be equal. Thus, having intellectual tools does not guarantee faultless reasoning. One must also decide to use them.

Notation and Conventions

1.5. Logic and Set Theory. It is assumed that the student of this book is knowledgeable in basic logical reasoning and set theory, the notation of which is briefly recapped here and used throughout. We write $\subseteq$ for the inclusion of sets, and $\subset$ for the proper (or strict) inclusion of sets. Thus, for any set X, we have $X \subseteq X$, $X \not\subset X$, and $\emptyset \subseteq X$ where $\emptyset$ is the empty set. As usual, $\cup$ and $\cap$ stand for the union and the intersection of sets, respectively. We may sometimes denote the union of disjoint sets by $+$ or by the summation sign $\sum$. The union of all the sets in a family $\mathcal{F}$ of subsets is symbolized by

$$\cup\mathcal{F} = \{x \mid x \in Y \text{ for some } Y \in \mathcal{F}\}, \tag{1.2}$$

and the intersection of all those sets by

$$\cap\mathcal{F} = \{x \mid x \in Y \text{ for all } Y \in \mathcal{F}\}. \tag{1.3}$$

The *complement* of a set Y with respect to some fixed ground set X including Y is the set $\overline{Y} = X \setminus Y$. The set of all the subsets, or *power set*, of a set X is denoted by 2^X. From (1.2) and (1.3), we get $\cup 2^X = X$ and $\cap 2^X = \emptyset$ (because $\emptyset \in 2^X$). The size (or cardinality, or cardinal number) of a set X is written as $|X|$. The *Cartesian product* of two sets X and Y is defined as

$$X \times Y = \{(x, y) \mid x \in X \,\&\, y \in Y\}$$

where (x, y) denotes an ordered pair and & means the logical connective 'and'; writing $\Leftrightarrow$ for 'if and only if,' we thus have

$$(x, y) = (x', y') \quad \Longleftrightarrow \quad (x = x' \;\&\; y = y').$$

More generally, $(x_1, \dots, x_n)$ denotes the ordered n-tuple of the elements $x_1, \dots, x_n$, and we have

$$X_1 \times \dots \times X_n = \{(x_1, \dots, x_n) \mid x_1 \in X_1, \dots, x_n \in X_n\}.$$

The symbols $\mathbb{N}$, $\mathbb{Q}$, and $\mathbb{R}$ stand for the sets of natural numbers, rational numbers, and real numbers, respectively; $\mathbb{R}_+$ denotes the set of nonnegative real numbers.

1.6. Warning. An expression such as $A = \{x, x, y, z, w\}$ is an instance of improper writing. It certainly (!) does not mean that the set A contains two copies of x. The best interpretation that could be given is:

$$\{x, x, y, z, w\} = \{x, y, z, w\}$$

which is consistent with the Axiom of Extensionality: TWO SETS ARE EQUAL IF AND ONLY IF THEY CONTAIN THE SAME ELEMENTS.

1.7. Binary Relations, Partitions. A set R is a *binary relation* if there are two sets X and Y such that

$$R \subseteq X \times Y. \tag{1.4}$$

Thus, a binary relation is a set of ordered pairs $xy \in X \times Y$, where xy is an abbreviation of (x, y). In such a case, we often write xRy to mean $xy \in R$. If $X = Y$ in (1.4), then R is said to be a binary relation *on* X. A binary relation is a *quasi order* on a set X if it is *reflexive* and *transitive* on X, that is, for all x, y, and z in X,

$$xRx \qquad\qquad\qquad\qquad \text{(reflexivity)}$$
$$xRy \;\&\; yRz \quad \Rightarrow \quad xRz \qquad\qquad \text{(transitivity)}$$

(where '$\Rightarrow$' means 'implies' or 'only if'). A quasi order R is a *partial order* on X if it is *antisymmetric* on X, that is, for all x and y in X

$$xRy \;\&\; yRx \quad \Rightarrow \quad x = y.$$

A binary relation R is an *equivalence relation* on a set X if it is reflexive, transitive, and *symmetric* on X, that is, for all x, y in X, we have

$$xRy \quad \Longleftrightarrow \quad yRx.$$

The following construction is standard. Let R be a quasi order on a set X. Define the relation $\sim$ on X by the equivalence

$$x \sim y \quad \Longleftrightarrow \quad (xRy \ \& \ yRx). \tag{1.5}$$

It is easily seen that $\sim$ is reflexive, transitive, and symmetric on X, that is, $\sim$ is an equivalence relation on X (cf. Problems 13 and 14). For any x in X, define the set $[x] = \{y \in X \mid x \sim y\}$. The family $\mathcal{X} = \{[x] \mid x \in X\}$ of subsets of X is called a *partition* of X *induced* by $\sim$. Any partition $\mathcal{X}$ of X satisfies the following three properties

[1] $Y \in \mathcal{X}$ implies $Y \neq \emptyset$;

[2] $Y, Z \in \mathcal{X}$ and $Y \neq Z$ imply $Y \cap Z = \emptyset$;

[3] $\cup \mathcal{X} = X$.

Conversely, any family $\mathcal{X}$ of subsets of a set X satisfying [1], [2], and [3] is called a partition of X.

1.8. Finite Induction. We take it for granted that the reader is conversant with finite induction. For example, to establish by induction the formula

$$1 + 2 + \ldots + n = \frac{n(n+1)}{2}, \tag{1.6}$$

for all $n \in \mathbb{N}$, we can first verify it for $n = 1$, and then prove that if it holds for $n = k$ (for any arbitrarily chosen $k \in \mathbb{N}$), then it must also hold for $n = k + 1$. For instance, Eq. (1.6) is verified for $n = 1$ because $1 = \frac{1 \cdot 2}{2}$, and if it also holds for $n = k$, then

$$1 + 2 + \ldots + k + (k+1) = \frac{k(k+1)}{2} + (k+1) = \frac{(k+1)(k+2)}{2}$$

by simple algebra, and so (1.6) also holds for $n = k + 1$. It holds thus for all $n \in \mathbb{N}$. Problems 12(a), (b), and (c) at the end of this chapter are intended to help the reader, if need be, to refresh his or her mastery of the finite induction technique of proof.

1.9. Organization of this Book. The chapters of this book are divided into sections, and the sections into so-called paragraphs, which often have titles such as 'Definition,' 'Remarks,' 'Theorem,' or 'Example.' Only the paragraphs are numbered, using a single alphanumeric system, as in '**1.9. Organization of this Book**' just

above. As exemplified in paragraph 1.5 on page 7 by the terms 'complement,' 'power set,' and 'Cartesian product,' defined terms are
indicated by slanted font. Occasionally, a section or a paragraph's
title may be 'starred' by using the symbol '∗.' This indicates that
its content may be skipped at first reading. The formulas are also
numbered alphanumerically, as in Eq. (1.6) on p. 9. We always use
parentheses in such cases. For convenience, we give on page 267 a
glossary of frequently used symbols.

Problems

1. Explain why the data of Example 1.2 illustrate the conjunction fallacy.
   Define each of the relevant events precisely.

2. Complete the following equation by providing a definition by extension of
   the set in the left-hand side: $\{n \in \mathbb{N} \mid 1 < |n - 3|\} \cap \{1, 2, 7\} = $.

3. Using logical or mathematical arguments rather than Venn diagrams, verify the two equalities below, which appear in the proof of Theorem 3.2.
   Note that the complement of some set A is defined with respect to some
   (fixed) ground set Ω.

   (a) $A \cap B = \overline{(\overline{A} \cup \overline{B})}$ for any $A, B \subseteq \Omega$.

   (b) $A \setminus B = \overline{(B \cup \overline{A})}$ for any $A, B \subseteq \Omega$.

4. Using the Axiom of Extensionality (Two sets are equal if and only if they
   contain the same elements), prove that there is at most one empty set.

5. Let Ω be the ground set, with A, B, and C subsets of Ω. Prove that:

   (a) $A \subseteq B \Leftrightarrow \overline{A} \cup B = \Omega$;

   (b) $A \subseteq B \Leftrightarrow \overline{B} \subseteq \overline{A}$;

   (c) $A \subseteq B \Leftrightarrow (A \cap \overline{B}) \subseteq (C \cap \overline{C})$.

6. For two sets X and Y, is it true that $X \times Y = Y \times X$ implies $X = Y$?
   Formulate a condition C such that the following assertion holds:
   If X and Y satisfy Condition C, then $X = Y \Longleftrightarrow X \times Y = Y \times X$.
   Prove the resulting assertion.

7. Complete the following equations and prove the results: (a) $\cup\{\{a\}\} = $;
   (b) $\cup 2^{\{\{a\}\}} = $; (c) $\cap 2^{\{\{a\}\}} = $; (d) $2^\emptyset = $; (e) $2^{\{\emptyset\}} = $.

8. What is a simple expression for the set
   $\{x \in \mathbb{R} \mid x^2 < 1\} \cap \{x \in \mathbb{R} \mid |x - 1| < |x + 2|\}$?

9. Prove that the relation $\sim$ defined by (1.5) is reflexive, symmetric, and transitive on X. Verify that the family $\mathcal{X} = \{[x] \,|\, x \in X\}$ satisfies Properties [1], [2] and [3].

10. Is it true that $A \subseteq B$ iff $2^A \subseteq 2^B$? Prove your response.

11. For a finite set X, what is the value of $2^{|X|}$? Try to prove your assertion.

12. For each of the following, provide a proof by induction:

    (a) $1^2 + 2^2 + 3^2 + \ldots + n^2 = \frac{n(n+1)(2n+1)}{6}$.

    (b) $1 \cdot 2 + 2 \cdot 3 + 3 \cdot 4 + \ldots + n(n+1) = \frac{n(n+1)(n+2)}{3}$.

    (c) With $0 \le x < 1$, we have

    $$1 + x + x^2 + \ldots + x^n = \frac{1 - x^{n+1}}{1 - x},$$

    and so $\sum_{n=0}^{\infty} x^n = (1 - x)^{-1}$.

    (d) $5^n - 1$ is divisible by 4 for $n = 1, 2, \ldots$.

    (e) $6 \cdot 7^n - 2 \cdot 3^n$ is divisible by 6 for $n = 1, 2, \ldots$.

13. Let $\mathcal{X}$ be a partition of some set X in the sense of Def. 1.7. For all $x \in X$, define $[x]$ as the class of the partition $\mathcal{X}$ that contains x. Define also a relation V on X by: $xVy \Leftrightarrow [x] = [y]$. Prove that V is an equivalence relation on X. (In fact, $\mathcal{X}$ is the partition induced by V.)

14. Prove in detail that the inclusion relation $\subseteq$ is reflexive, transitive and antisymmetric. (These three conditions define a partial order; cf. 1.7.)

Chapter 2

Sample Space and Events

Probability Theory is the part of mathematics concerned with the analysis of random (or chance) experiments. An experiment is called *random* if its outcome cannot be predicted with certainty[1]; it is also required that, if the experiment is repeated under the same conditions, the possible outcomes occur with a certain statistical regularity. A better grasp of this concept will emerge from considering a few examples.

2.1. Example. (COIN TOSSING.) A coin (not necessarily fair) is tossed twice. A typical outcome of this experiment would be represented by an ordered pair such as (H, T), where 'H' stands for 'head' and 'T' stands for 'tail.' The set of all possible outcomes of this experiment is thus represented by

$$\{(H, H), (H, T), (T, H), (T, T)\}.$$

If this experiment is repeated over and over again, with a fair coin, the outcome (H, T) will occur in about 25% of all cases. This illustrates the concept of 'statistical regularity' mentioned in the introductory paragraph. Providing a more precise elaboration of this concept is one of the aims of Probability Theory.

2.2. Example. (COIN TOSSING UNTIL A HEAD APPEARS.) In the same situation, the coin is tossed until a head appears. The experiment records the number of trials required. An outcome is thus represented here by a natural number k, which symbolizes the fact

---

[1]From the standpoint of a particular observer. Randomness may depend on one's point of view. An experiment may be random for one observer, but not for another one who would be aware of some deterministic mechanism generating the outcome.

that the first head appeared on the kth trial. The set of all possible outcomes can thus be represented by the set of natural numbers $\mathbb{N} = \{1, 2, ..., k, ...\}$.

These two examples illustrate a fundamental concept of Probability Theory: THE SAMPLE SPACE IS A SET OF DESCRIPTIONS OF ALL THE POSSIBLE OUTCOMES OF AN EXPERIMENT.

Thus, the sample spaces of Examples 2.1 and 2.2 are, respectively,

$$\{(H, H), (H, T), (T, H), (T, T)\} \quad \text{and} \quad \{1, 2, \ldots, k, \ldots\}.$$

For simplicity, we shall identify 'outcome' and 'description' (of an outcome). We shall consider a few additional examples.

2.3. Example. (LEARNING.) On each trial of a learning experiment, a hungry animal, a rat say, is placed at the starting point of a T-maze (i.e., a T-shaped maze; see Fig. 2.1). The rat runs to the choice point and then turns right or left (retracing is not allowed). On every trial a pellet of food is placed in the box on the right side. The animal eventually learns to turn to the right and find the food.

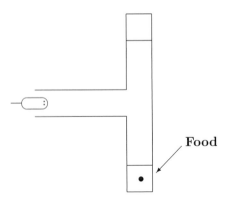

Figure 2.1: The T-maze in the learning experiment.

The experimenter records the choice of the animal on every trial. A typical outcome of this experiment is a succession of left and right turns, and can be described by a sequence such as

$$(L, L, R, L, R, R, \ldots, R, R, R, R, R, R, \ldots).$$

In this particular case, the animal has turned left on the first and second trial, right on the third trial, and so on. The sample space of this experiment is the set of all such sequences.

2.4. Remark. This choice of a sample space involves some idealization: In practice, the length of any sequence of choices produced by a particular animal will never exceed a few hundred trials. A priori, one may be tempted to consider as possible outcomes only those sequences of 1000 trials or less (say). In fact, there are solid theoretical reasons to take as our sample space for the T-maze experiment the set of all infinite sequences of left and right turns. This remarks applies to many probabilistic experiments based on a similar pattern. We shall come back to this point in the next section.

2.5. Example. (SIMPLE REACTION TIME.) A trial of this experiment begins with the presentation of a warning signal followed, after a variable delay, by a fixed visual stimulus. The subject is required to react as fast as possible to the stimulus by pressing a key. The latency of the response is recorded. Possible sample spaces for this experiment would be the set $]0, \infty[$ of all positive real numbers or the set $\mathbb{R}$ of all real numbers.[2] This example illustrates a large class of situations in the behavioral, biological, and physical sciences in which a single numerical measure of a phenomenon is recorded.

The Sample Space

Our first basic notion is that of the *sample space* as a set of descriptions of all the possible *outcomes* of an experiment. 'Sample space' and 'outcome' are technical terms. The term 'possible' deserves some comments. In Example 2.1, we took as our sample space the set

$$\{(H, H), (H, T), (T, H), (T, T)\}, \tag{2.1}$$

deliberately omitting the case where the coin falls on its edge. Another possibility for the sample space of this experiment would have been, with 'E' representing 'edge':

$$\{(H, H), (H, T), (T, H), (T, T), (H, E), (T, E),$$
$$(E, H), (E, T), (E, E)\}. \tag{2.2}$$

---

[2]One can argue that $\mathbb{R}$ would be a better choice: Choosing $]0, \infty[$ as a sample space forces the researcher to delete from the data all the anticipatory responses yielding negative reaction times. Such responses may be regarded by some researchers as revealing important aspects of the phenomenon under study.

Actually, both (2.1) and (2.2) would be appropriate sample spaces. A coin falling on its edge is such a rare event that the set given in (2.1) would, for all practical purposes, be quite adequate. A better reason for adopting (2.1) would arise if the experimenter is following the procedure of discarding any result which cannot be described by H or T. But (2.2) would also be acceptable. WHAT IS CRITICAL IS THAT ALL FEASIBLE OUTCOMES HAVE A DESCRIPTION IN THE SAMPLE SPACE. IT IS OF NO IMPORTANCE THAT THE SAMPLE SPACE CONTAINS OUTCOMES THAT NEVER HAPPEN IN PRACTICE: IN THE FRAMEWORK OF A PROBABILISTIC MODEL, SUCH OUTCOMES MAY BE ASSIGNED PROBABILITY ZERO.

In Example 2.2 (the toss of a coin until a head appears), an appropriate sample space is the set of natural numbers. Each natural number is a representation of a possible outcome. A particular element k in $\mathbb{N}$ represents the outcome: The first head appears after $k - 1$ consecutive tails. In the reaction time experiment of Example 2.5, the standard convention is to take as the sample space the set $\mathbb{R}$ of all real numbers, even though, in most experimental conditions, it may be unreasonable to assume that negative reaction times (which may be interpreted as anticipations), or very large reaction times, say of the order of several hours, will occur.

Finally, let us turn to the learning experiment of Example 2.3, which is conceptually more difficult. As indicated, an outcome in this experiment is represented by a sequence such as

$$(L, L, R, L, R, R, \dots, R, R, R, R, R, R, \dots).$$

An appropriate sample space is thus the set of all such sequences. This is a fairly large set. (Actually, it is uncountable; cf. Problem 1.)

A sample space can thus be finite (as in Example 2.1), countable (as in Example 2.2), or uncountable (as in Examples 2.3 and 2.5).

2.6. Definition. A sample space which is either finite or countable is called *discrete*. Thus the sample spaces of Examples 2.1 and 2.2 are discrete.

In analyzing a random experiment, the first task is often to specify the sample space. In many cases, this may be easy to do. In some situations, however, the construction of the sample space may be challenging, even for finite sample spaces. A good example of such a more difficult case is given below.

2.7. Example. (THE SAMPLE SPACE IN THE GAME SHOW.) In the game show mentioned in Chapter 1, the participants were asked to open one of three doors, marked A, B, and C. Behind one of the doors was a valuable prize (for example, a car). A special feature of the game was that, after a contestant had made his or her choice of a door, the show host would open one of the two remaining doors and show that there was nothing of value behind it. The contestant was then given the option to switch. For example, suppose that the contestant chooses door A and the host opens door B. Should the contestant switch to door C? How should we analyze this situation from the viewpoint of a random phenomenon? A full description of an outcome consists in the following items:

1. Placement of the car behind a door. For the sake of concreteness, suppose that the car is placed behind door A.

2. Choice of a door by the contestant, say door B.

3. Opening of a door by the show host. In this case, this door is necessarily door C.

We adopt the following representation for such outcomes: the door hiding the car will be marked with a star; the door chosen by the contestant will be underlined; and the door opened by the host, revealing a goat, will be marked by the dagger symbol '†.' For the particular outcome described in (1)-(3) above, we have thus the symbolic representation: $A^* \underline{B} C^\dagger$.

Notice that, in this particular case, the host had no choice: Door C was the only one available. However, if the contestant had chosen the correct door A, then the host would have had the choice between the two doors B and C. Thus, two possible outcomes arise from a case where the contestant chooses the correct door. The complete list of the possible outcomes is given in Table 2.1.

Table 2.1: List of outcomes in the game show

$\underline{A}^* B^\dagger C$	$\underline{A} B^* C^\dagger$	$\underline{A} B^\dagger C^*$
$\underline{A}^* B\ C^\dagger$	$A^\dagger \underline{B}^* C$	$A^\dagger \underline{B}\ C^*$
$A^* \underline{B}\ C^\dagger$	$A\underline{B}^* C^\dagger$	$A^\dagger B\ \underline{C}^*$
$A^* B^\dagger \underline{C}$	$A^\dagger B^* \underline{C}$	$AB^\dagger \underline{C}^*$

In this situation, the sample space contains 12 outcomes.[3] We shall pursue the analysis of this example in Chapter 4 and introduce probabilities.

2.8. Example. (A STANDARD EXAMPLE.) An important sample space, frequently used in scientific applications, is the Cartesian product

$$\mathbb{R}^n = \underbrace{\mathbb{R} \times \ldots \times \mathbb{R}}_{n \text{ times}}, \tag{2.3}$$

with outcomes of the form of real vectors $\mathbf{x} = (x_1, \ldots, x_n)$. An example is offered by an experiment in which various numerical characteristics of human subjects—such as blood pressure, weight, and height—are simultaneoulsy measured. Thus, each subject would corresponds to a triple (b, w, h), where b, w, and h represents his or her blood pressure, weight, and height, respectively. Each triple (b, w, h) is an outcome and $\mathbb{R}^3$ is the sample space.

The Concept of an Event

Our second basic notion is that of an 'event.' The fundamental idea is that an event is represented by a subset of the sample space. Again, we begin by considering examples.

1. In Example 2.1, the event 'observing at least one head' is represented by the set

$$\{(H, T), (T, H), (H, H)\}.$$

2. In Example 2.2, the event 'obtaining a head within 20 tosses' is denoted by
$$\{k \in \mathbb{N} \mid k \leq 20\}.$$

The next example may be harder to digest.

3. In Example 2.3, the event 'the animal turns left at the 15th trial' may be represented by the set

$$\{s \mid s \text{ is a sequence in } \{L, R\} \text{ and } s_{15} = L\}.$$

---

[3]We could have included the contestant's decision in the description of an outcome. If we had done so, the sample space would have contained 24 points. Each of the outcomes in Table 2.1 would have generated 2 cases: (1) stay and (2) switch. We shall see that there is no need for such a complication.

(We recall that a *sequence* in a set S is a function $f : \mathbb{N} \to S$ mapping the set $\mathbb{N}$ of natural numbers into S. The standard convention for sequences is to write f_n rather than $f(n)$ to denote the value of the function f at the point n.)

4. In the game show of Example 2.7, consider the statement 'the contestant switches and loses.' This statement describes an event in the sample space specified by Table 2.1. For example, the outcome $A\underline{B}^*C^\dagger$ is a point of that event: The contestant chooses the correct door B, is shown the goat behind door C, switches to door A, and loses. Thus, the event 'the contestant switches and loses' contains all the outcomes corresponding to cases in which the initial choice was correct. This event is thus represented by the set:

$$\{\underline{A}^*B^\dagger C, \underline{A}^*BC^\dagger, A^\dagger \underline{B}^*C, A\underline{B}^*C^\dagger, A^\dagger B\underline{C}^*, AB^\dagger \underline{C}^*\}.$$

5. In the sample space $\mathbb{R}^3$ discussed in Example 2.8, all the open sets of the form $]b_1, b_2[\times]w_1, w_2[\times]h_1, h_2[$ are events.[4]

IN GENERAL: EACH EVENT PERTAINING TO A PARTICULAR EXPERIMENT MAY BE IDENTIFIED WITH A SUBSET OF THE RELEVANT SAMPLE SPACE.

A natural question is: Should the converse also hold? That is, should we consider each subset of the sample space as an event? This is not necessarily so. With finite sample spaces, the intuitively appealing position that each subset of the sample space defines an event is reasonable and could be adopted. With infinite sample spaces, however, intuition fails to be a dependable guide.

FOR TECHNICAL REASONS, THE GENERAL ASSUMPTION THAT ALL SUBSETS OF THE SAMPLE SPACE ARE EVENTS TO WHICH A PROBABILITY COULD BE ASSIGNED WOULD CREATE MATHEMATICAL DIFFICULTIES. FURTHER COMMENTS WILL BE MADE ON THIS ISSUE LATER ON.

---

[4] As an example, consider the event of observing for some individual a blood pressure strictly between 140 and 160 (mm Hg), a weight strictly between 80 and 90 (kg), and a height strictly between 1.80 and 1.90 (m).

Indicator Functions *

When dealing with sets, it is occasionally useful to perform computations and to apply the operations of arithmetic. A handy device in this connection is introduced in the next definition.

2.9. Definition. Let $\mathcal{F}$ be a family of sets on a set Ω. The *indicator (function)* of any set $Y \in \mathcal{F}$ is the function $I_Y : \Omega \to \{0, 1\}$ defined by the formula

$$I_Y(y) = \begin{cases} 1 & \text{if } y \in Y, \\ 0 & \text{if } y \in \Omega \setminus Y. \end{cases}$$

The basic set theoretical operations are readily translated in the language of indicator functions. For instance, it is easily verified that if I_X and I_Y are the indicator functions of the sets X and Y, respectively, then

$$I_{X \cap Y} = I_X \cdot I_Y \qquad (2.4)$$
$$I_{X \cup Y} = I_X + I_Y - I_{X \cap Y} \qquad (2.5)$$

and if $X \subseteq Y$, then

$$I_X \leq I_Y. \qquad (2.6)$$

The proofs of these results are left as Problem 8.

Problems

1. In the context of the learning experiment of Example 2.3, argue that the set of all sequences of left-right choices made by the rat is uncountable, specifically, is as large as the set of all real numbers. (Hint: Think about the binary notation of the real numbers.)

2. Consider a variant of the game show of Example 2.7 in which the contestant must choose one of four doors, with one hiding an expensive prize and the remaining three nothing of value. How many points are in the sample space? Describe a few outcomes and events precisely.

3. For each of the following, construct an appropriate sample space and define one or two events. If you find that one of the descriptions is unclear, specify it as you see fit. (Resolving the ambiguity is part of the problem.)

   (a) There are three hats on the table: two green hats and one red hat. There are two ties in the closet: one green tie and one red tie. Each of two brothers, Ralph and Rudolph, picks one hat and one tie at random.

(b) Eight swimmers compete in the final of an Olympic swimming event, the fastest three of whom receive medals—gold, silver, and bronze. (Only those receiving medals should be considered.)

(c) A subject is given a list of twenty words to study and is later asked to repeat the words in order. (This is called a *serial recall* task.)

(d) Dr. Ofthwahl, an anthropologist, wishes to verify his theory that the length/width ratio of the face of any adult male of the Arumbaya tribe equals approximately 1.5. With this aim, he collects photographs of the faces of dozens of subjects and takes measurements.

(e) Each of n letters is removed from its envelope. The letters are thoroughly mixed and then placed blindly back into the envelopes, with each envelope getting exactly one letter.

(f) A gambler plays a game in which she either wins \$1 or loses \$1 on each turn. She begins the game with \$$k$ and wishes to know how long she will be able to play before either her money runs out or she accumulates \$$N$.

4. Suppose four balls are distributed at random into three boxes. How many points does the sample space have? Can you find a general formula (for n balls in m boxes)?

5. A thin, straight rod of unit length is broken in two places at random, and a value is sought for the probability that a triangle may be formed from the resulting pieces. What is the sample space? How large is that sample space? (Cf. Beckmann, 1967.)

6. In a voting scheme called *approval voting*, a participant may vote for any subset of the candidates. Suppose that there is a set $C = \{c_1, \dots, c_n\}$ of candidates.

   (a) What is the sample space for such a scheme? What is the size of this sample space?

   (b) How many possible events are there for this sample space?

   (c) Define the event 'candidate c_1 is not voted for by the participant.' Write out this event explicitly for the case $n = 3$.

7. In the learning experiment of Example 2.3, suppose that the experimenter is also recording the time required for the animal to complete a trial, i.e., to find either of the left or the right boxes. What would be the sample space then?

*8. Prove Formulas (2.5), (2.4), and (2.6). Also, verify that if $\overline{X}$ is the complement of an event X in some sample space, then $I_X + I_{\overline{X}} = I_\Omega = 1$.

Chapter 3

Probability and Area

For the moment, it is sufficient to remember that the collection of all events is represented by some family $\mathcal{F}$ of subsets of the sample space, which we denote by the Greek letter Ω (upper case 'omega'). We shall identify each outcome with its representing point in Ω, and each event with its representing subset of Ω. Not all families of subsets are acceptable, however. We shall introduce some intuitively reasonable constraints on $\mathcal{F}$.

Suppose that A and B are events. Thus, $A, B \in \mathcal{F}$. It is natural to require that $A \cup B$ is also a member of $\mathcal{F}$ since $A \cup B$ would be the event that is realized when at least one of the two events A, B occurs, that is, whenever an outcome arises which belongs to $A \cup B$. In the same vein, if $A \in \mathcal{F}$, we should also have $(\Omega \setminus A) \in \mathcal{F}$, since $\Omega \setminus A$ would be the event that occurs exactly when A does not occur. In any discussion in probability, we often have in mind a fixed sample space Ω. This justifies the following abbreviation. For any event A, we shall denote by $\overline{A}$ the complement of A with respect to Ω, that is, $\overline{A} = \Omega \setminus A$. The key concept introduced by these considerations will now be defined.

Axioms for a Field

3.1. Definition. Let Ω be a set. Then $\mathcal{F}$ is a *field* of sets on Ω if $\mathcal{F}$ is a non empty family of subsets of Ω satisfying, for all $A, B \in \mathcal{F}$,

[F1] $A \cup B \in \mathcal{F}$;

[F2] $\overline{A} \in \mathcal{F}$.

23

In the language of Probability Theory, Axioms [F1] and [F2] state that if A and B are events, then both $A \cup B$ and $\overline{A}$ are events (and, by symmetry, also $\overline{B}$). In mathematics, [F1] and [F2] are sometimes expressed by stating that $\mathcal{F}$ is *closed under (finite) union* and *closed under complementation*. This terminology will occasionally be used in the sequel. As we shall see, a consequence of this definition is that if both A and B are events, then $A \cap B$, which represents the joint realization of A and B, is also an event. To express this fact, we shall say that the field $\mathcal{F}$ of sets is *closed under intersection*. Some consequences of Definition 3.1 are expressed in the theorem below.

Basic Results

3.2. Theorem. *If $\mathcal{F}$ is a field of sets on a set Ω, then*

(i) $\Omega \in \mathcal{F}$;

(ii) $\emptyset \in \mathcal{F}$;

(iii) $A \setminus B \in \mathcal{F}$, *for all* $A, B \in \mathcal{F}$.

   Moreover, if $A_1, A_2, \ldots, A_n \in \mathcal{F}$, *then*

(iv) $\cup_{i=1}^{n} A_i \in \mathcal{F}$;

(v) $\cap_{i=1}^{n} A_i \in \mathcal{F}$.

   In (iv), the notation $\cup_{i=1}^{n} A_i$ symbolizes the set containing all those points which are in at least one of the sets A_i, for $i = 1, 2, \ldots, n$. Similarly, in (v), $\cap_{i=1}^{n} A_i$ means the set containing all those points contained in all the sets A_i.

   PROOF. (i) We recall that $\mathcal{F}$ is not empty. Thus, there exists some set $A \in \mathcal{F}$. Since $\mathcal{F}$ is closed under complementation, we also have $\overline{A} \in \mathcal{F}$. But $\mathcal{F}$ is also closed under union. We conclude that $\Omega = A \cup \overline{A} \in \mathcal{F}$.

   (ii) This follows readily from (i) and the fact that $\emptyset = \overline{\Omega} \in \mathcal{F}$ since $\mathcal{F}$ is closed under complementation.

   (iii) From set theory, it is easily shown that

$$A \setminus B = \overline{(B \cup \overline{A})}. \qquad (3.1)$$

(The verification of this fact is left to the student; cf. Problem 3 in Chapter 1.) Since $A \in \mathcal{F}$, and $\mathcal{F}$ is closed under complementation, we have $\overline{A} \in \mathcal{F}$. This result, together with $B \in \mathcal{F}$ and the fact that $\mathcal{F}$ is closed under union, implies that $(B \cup \overline{A}) \in \mathcal{F}$. Using again the closure of $\mathcal{F}$ under complementation, we derive that the r.h.s.[1] of the equality (3.1) is a member of $\mathcal{F}$, establishing (iii).

(iv) We prove this result by induction on n. For $n = 1$, we clearly have $\cup_{i=1}^{1} A_i = A_1 \in \mathcal{F}$. Suppose that the result holds for $n = m$, and consider the case $n = m + 1$. Then, by definition of the union of the sets $A_1, A_2, \ldots, A_{m+1}$,

$$\cup_{i=1}^{m+1} A_i = (\cup_{i=1}^{m} A_i) \cup A_{m+1}. \tag{3.2}$$

By assumption, the sets $A_1, A_2, \ldots, A_{m+1} \in \mathcal{F}$. The induction hypothesis implies that we have $\cup_{i=1}^{m} A_i \in \mathcal{F}$. Thus, from (3.2) and the fact that $\mathcal{F}$ is closed under union, we derive that $\cup_{i=1}^{m+1} A_i \in \mathcal{F}$.

(v) We first prove that $A \cap B \in \mathcal{F}$ whenever $A, B \in \mathcal{F}$. This follows immediately from the conditions defining a field of sets and from the fact that

$$A \cap B = \overline{(\overline{A} \cup \overline{B})}, \tag{3.3}$$

which is a simple result from set theory. (The student is advised to verify this; see Problem 3a in Chapter 1.) Then, (v) is obtained by induction. The details are left to the student.

We have established parts (i) to (v) of the Theorem. The proof is thus complete. $\square$

3.3. Remark. Our strategy will be as follows. We shall start with a field of sets $\mathcal{F}$ (the sets of events) on a set Ω (the sample space). Next, we introduce the concept of a 'probability measure' $\mathbb{P}$ as a function assigning a number $\mathbb{P}(A)$, with $0 \leq \mathbb{P}(A) \leq 1$, to every event A in $\mathcal{F}$. Moreover, this function $\mathbb{P} : \mathcal{F} \to [0, 1]$ will be assumed to satisfy a number of axioms, a list of which is given in Chapter 4 (see 4.1). The probability measure $\mathbb{P}$ is not assumed in general to be defined on all subsets of the sample space. The reason for this is of a technical nature: In the case of an uncountable sample space Ω, the assumption that there exists a probability measure $\mathbb{P}$ defined on all subsets A of Ω leads to a contradiction. The details of the argument are beyond the scope of this course.

---

[1]We sometimes abbreviate 'left-hand side' and 'right hand side' (of a formula) by 'l.h.s.' and 'r.h.s.,' respectively.

Glossary of Terms

Table 3.1 gathers some everyday language expressions pertaining to events in Probability Theory, with the corresponding set theoretical formulas and concepts. The elements of this table are borrowed from Kolmogorov (1950, Chapter 1). We suppose that we have a sample space Ω, and a field $\mathcal{F}$ of subsets of Ω. The set I is the index set of the family $\{A_i\}_{i \in I}$ of events in $\mathcal{F}$. Notice that we have translated the equation $A = \Omega$ by the statement 'the event A is certain.' The axioms of probability are consistent with this language, and require that the sample space Ω have a probability equal to one. However, these axioms will also permit a situation in which $\mathbb{P}(A) = 1$ even though $A \neq \Omega$. This event A will then sometimes be said to be *of measure one*, without being certain. Similarly, we can have $\mathbb{P}(A) = 0$ for some non empty event A, which will then said to be *of measure zero*, without being impossible.

Table 3.1: Common Terms and Expressions in Probability Theory

Set Theory Formulas	Probability Theory
$A \in \mathcal{F}$	A is an event
$A, B \in \mathcal{F}, \quad A \cap B = \emptyset$	The events A and B are *incompatible*
$A = \emptyset$	The event A is *impossible*
$A = \Omega$	The event A is *certain*
$A_i \in \mathcal{F}$, for all $i \in I$, $\quad \cup_{i \in I} A_i = B$	The event B is realized when at least one of the events A_i is realized
$A_i \in \mathcal{F}$, for all $i \in I$, $\quad \cap_{i \in I} A_i = B$	The event B occurs when all of the events A_i are realized
$A, B \in \mathcal{F}, \quad A \subseteq B$	If event A occurs, then event B necessarily occurs
$A \in \mathcal{F}, \mathbb{P}(A) = 0$	The event A has measure zero

Geometrical Interpretation

A sound intuition for the concept of a probability measure emerges from considering each event A as a geometrical region, say, a region in the plane. The probability $\mathbb{P}(A)$ of the event A can be regarded as a measure of the area of the region of the plane representing the event A. In particular, the measure $\mathbb{P}(\Omega)$ of the area representing the sample space (the certain event) will be set equal to one. To illustrate this idea, we consider the case of three events A, B and C which are represented in Fig. 3.1. The left oval surrounds the region associated with event A, the upper right oval the region associated with event B, and the lower oval that associated with event C. The rectangular region surrounded by the frame of the figure corresponds to the sample space Ω. In this representation, $\mathbb{P}(A)$ is a measure of the region delimited by the left oval in Fig. 3.1. Similarly, $\mathbb{P}(A \setminus C)$ is a measure of the shaded area. The area of the frame has measure $\mathbb{P}(\Omega) = 1$.

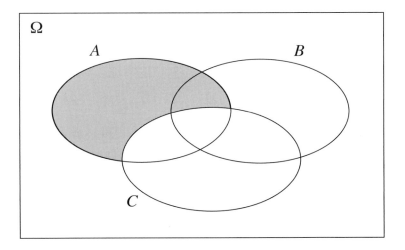

Figure 3.1: The oval regions corresponding to the events A, B, and C. The shaded area has measure $\mathbb{P}(A \setminus C)$.

Infinite Sample Spaces and σ-Fields

The definition of the family of all events as a field of sets in the sense of Def. 3.1 is adequate for finite sample spaces. However, the Axiom of 'Closure under Finite Union' (Condition [F1] in Def. 3.1) needs to be strengthened in the case of an infinite sample space (with an infinite collection of events). Suppose, for example, that you are in charge of the quality control of a piece of equipment, say, a battery, and that you are evaluating the duration of its performing

adequately. Specifically, you are concerned with the possibility that
a battery would fail before time t. It makes sense to consider the
time of failure as an outcome of a random experiment, with the
open interval $]0, \infty[$ as its sample space. Thus, testing a particular
battery corresponds to selecting a point in the sample space. The
event 'the battery fails before time t' is then represented by the open
interval $]0, t[$ (with $t > 0$). By symmetry, it is reasonable to require
that all the sets $]t, \infty[$ also be events. The Axiom of Closure under
Complementation implies that all the sets

$$[t, \infty[\; = \;]0, \infty[\; \setminus \;]0, t[$$

are events. Notice that

$$]t, \infty[\; = \; \cup_{n=1}^{\infty} [t + \frac{1}{n}, \infty[,$$

which specifically expresses the event $]t, \infty[$ as a countable union
of other events. This illustrates and motivates the axiom in the
following definition.

3.4. Definition. A field $(\Omega, \mathcal{F})$ is called a *σ-field* if it is *closed
under countable union,* that is, if for any countable collection A_1,
$A_2, \ldots, A_n, \ldots$ of sets in $\mathcal{F}$, we have $\cup_{n=1}^{\infty} A_n \in \mathcal{F}$.

Note that for any sets $A_1, A_2, \ldots, A_n, \ldots$, we have

$$\cap_{n=1}^{\infty} A_n = \overline{\cup_{n=1}^{\infty} \overline{A_n}}. \qquad (3.4)$$

As a σ-field, $\mathcal{F}$ is closed under both countable union and complementation. We conclude from Eq. (3.4) that $\mathcal{F}$ is also closed under
countable intersection.

Borel Fields and Sets

3.5. Definition. We consider the important case where the sample
space is the set $\mathbb{R}$ of real numbers. We recall that a set S of real
numbers is an *open* set of $\mathbb{R}$ if for any $x \in S$, there exists some $\delta > 0$
such that whenever $|x - y| < \delta$, then $y \in S$. The standard family
of events for $\mathbb{R}$ is a distinguished σ-field $\mathcal{B}$ containing all the open
sets of $\mathbb{R}$. In fact, $\mathcal{B}$ is the 'smallest' σ-field containing these open
sets. (For the meaning of 'smallest' in this sentence, see Problems

3 and 4.) The collection $\mathcal{B}$ is called the *Borel field* of $\mathbb{R}$, and the events in $\mathcal{B}$ are referred to as the *Borel sets* of $\mathbb{R}$, after the French mathematician Emile Borel (1871-1956), who played a prominent role in the foundations of probability theory.

These definitions extend naturally to the case where the sample space is $\mathbb{R}^n$, with the outcomes being represented by all the vectors of the form $\boldsymbol{x} = (x_1, \ldots, x_n)$. The family of events is the smallest σ-field containing all the open sets of $\mathbb{R}^n$, where the concepts of 'open set' of $\mathbb{R}^n$ is defined in a manner similar to that given above in the case of the open sets of $\mathbb{R}$. (Namely, a subset S of $\mathbb{R}^n$ is *open* if whenever it contains some point $\boldsymbol{x}$, it also contains, for some $\delta > 0$, all those points $\boldsymbol{y} \in \mathbb{R}^n$ satisfying $|\boldsymbol{x} - \boldsymbol{y}| < \delta$, where $|.|$ denotes the Euclidean distance.) The events in the sample space $\mathbb{R}^n$ are also called *Borel sets*.

Problems

1. Let $\Omega = \{a, b, c, d, e\}$ be the sample space, and suppose that $\{a\}$ and $\{b, c\}$ are events. Define a field $\mathcal{F} \neq 2^\Omega$ consistent with that information.

2. Let $\mathbb{R}^2$ be the sample space. Using the axioms of a Borel field, show that all the sets listed below are Borel sets:

   (a) the closed rectangles of the form $[x_1, y_1] \times [x_2, y_2]$;

   (b) the sets $\{(x, y)\}$;

   (c) the set of all points of $\mathbb{R}^2$ which have rational coordinates;

   (d) any line with equation $ax + by + c = 0$, where a, b, and c are arbitrary real numbers.

   At this point, you may be developing the impression that any subset of $\mathbb{R}^n$ (for any natural number n) is a Borel set. This is far from true, but we shall not prove this fact here (see, for example, Royden, 1966).

3. Let $\mathcal{F}_1$ and $\mathcal{F}_2$ be two σ-fields on the same sample Ω. Prove that $\mathcal{F}_1 \cap \mathcal{F}_2$ is also a σ-field on Ω. Generalize this result: Suppose that $\mathcal{G}$ is a collection of σ-fields on the same sample space Ω. Prove that the intersection $\cap\mathcal{G}$ of all these σ-fields is itself a σ-field on Ω.

4. Keeping in mind the result of Problem 3, reflect on the concept of the smallest σ-field on $\mathbb{R}$ containing all the open sets of $\mathbb{R}$. Can you show that such a smallest σ-field always exists?

5. Check whether the field that you have defined in your solution of Problem 1 is the smallest field satisfying the required constraints. If it is not, define such a smallest field.

6. Let the sample space Ω be the set of all rankings of the three alternatives a_1, a_2, and a_3, with ties allowed. (Such rankings, called *(strict) weak orders*, arise for instance from polls requiring respondents to rate each of several choices on a numerical scale.) Find an appropriate coding of the outcomes, and give examples of nonempty events A, B, and C in Ω satisfying the following expressions:

   (a) $A \cup B \cup C = \Omega$, with $A \neq \Omega$, $B \neq \Omega$, and $C \neq \Omega$;

   (b) $A \subseteq \bar{B}$, $C \subseteq \bar{B}$, and $A \cap C \neq \emptyset$;

   (c) $A \subset B \subset C$ (so that $A \neq B$, $B \neq C$, and $A \neq C$);

   (d) Write the events: 'a_3 is ranked second' and 'a_2 and a_3 are tied.'

7. With Ω as in Problem 6, give a field of sets $\mathcal{F}$ on Ω such that $|\mathcal{F}| = 4$.

8. Show that the power set of any nonempty set is a σ-field.

9. In the experiment performed by Dr. Ofthwahl in Problem 3d of Chapter 2, the sample space can be taken to be $]0, \infty[$, the set of positive real numbers. The event 'the *length* : *width* ratio equals 1.5' is then represented by the set $L = \{(x, y) \in]0, \infty[\times]0, \infty[\mid \frac{x}{y} = 1.5\}$. Is L a Borel set? Why or why not?

10. Give an example of a field of sets on $\mathbb{R}$ which is not a σ-field.

11. A regular 6-face die is tossed three times. Find an appropriate notation for the outcomes of the sample space, and write explicitly the event 'the sum of the three tosses is 17.'

Chapter 4

Probability Measures

We shall define here the concept of a probability measure $\mathbb{P}$ as a function assigning a number $\mathbb{P}(A)$ to each event A in a field $\mathcal{F}$. As indicated in Chapter 3, this function $\mathbb{P}$ will be defined so as to satisfy, for all events A, the two conditions:

$$\mathbb{P}(\Omega) = 1, \qquad (4.1)$$
$$0 \le \mathbb{P}(A) \le 1, \qquad (4.2)$$

A third basic condition requires some thought. Suppose that A and B are two incompatible events (thus, $A \cap B = \emptyset$). In line with the geometrical interpretation of $\mathbb{P}(A)$ as measuring the area of a region representing the event A, it seems also sensible to demand that

$$\mathbb{P}(A \cup B) = \mathbb{P}(A) + \mathbb{P}(B). \qquad (4.3)$$

It is easily shown that (4.3) readily leads to the apparently more general condition: For pairwise incompatible events $A_1, A_2, \ldots, A_n$, we always have

$$\mathbb{P}(\cup_{i=1}^{n} A_i) = \sum_{i=1}^{n} \mathbb{P}(A_i).$$

This condition is often referred to as the 'finite additivity' of the probability measure. One can, however, be more demanding, and require that if $A_1, A_2, \ldots, A_i, \ldots$ is a (countably) infinite sequence of pairwise incompatible events, then $\bigcup_{i=1}^{\infty} A_i$ is an event, and moreover

$$\mathbb{P}(\cup_{i=1}^{\infty} A_i) = \sum_{i=1}^{\infty} \mathbb{P}(A_i). \qquad (4.4)$$

In this equation, $\cup_{i=1}^{\infty} A_i$ symbolizes the union of the countable collection of events $\{A_i \,|\, i \in \mathbb{N}\}$, and $\sum_{i=1}^{\infty} \mathbb{P}(A_i)$ represents the sum of a numerical series, a standard concept of calculus. A probability measure $\mathbb{P}$ satisfying (4.4) is said to be *countably additive* or *σ-additive*.

Most applications of probability require the assumption of countable additivity. Since, however, the basic concepts of probability may be introduced in the simple context of finite additivity, we begin by discussing this case.

SINCE COUNTABLE ADDITIVITY IMPLIES FINITE ADDITIVITY, ANY RESULT OR DEFINITION FORMULATED IN THE CONTEXT OF FINITE ADDITIVITY IS VALID IN THE COUNTABLE CASE.

Finitely Additive Probability Spaces

4.1. Definition. Let $\mathcal{F}$ be a field of sets on a set Ω. Then, the triple $(\Omega, \mathcal{F}, \mathbb{P})$ is a *finitely additive probability space* iff $\mathbb{P}$ is a real valued function on $\mathcal{F}$ satisfying the following three conditions: For all $A, B \in \mathcal{F}$,

[K1] $\mathbb{P}(\Omega) = 1$;

[K2] $\mathbb{P}(A) \geq 0$;

[K3] if $A \cap B = \emptyset$, then $\mathbb{P}(A \cup B) = \mathbb{P}(A) + \mathbb{P}(B)$.

The function $\mathbb{P}$ is called a *probability measure*. In the special case where the field $\mathcal{F}$ is a σ-field and moreover Eq. (4.4) holds for any countable family $\{A_i \,|\, i \in \mathbb{N}\}$ of pairwise disjoint events, then $(\Omega, \mathcal{F}, \mathbb{P})$ is called a *probability space*. Axioms [K1], [K2], and [K3] are often referred to as *Kolmogorov's Axioms* (Kolmogorov, 1950). We shall illustrate this definition by a few examples.

4.2. Example. Suppose that Ω is a finite set, and let $\mathcal{F}$ be a field of its subsets. For any event $A \in \mathcal{F}$, we define

$$\mathbb{P}(A) = \frac{|A|}{|\Omega|}.$$

In other terms, the value of the probability measure $\mathbb{P}$ for a particular event A is the ratio of the number of elements of A to the total number of elements in the sample space Ω. The student should verify that this definition does indeed specify a probability measure (see Problem 1). This probability measure is sometimes called the *counting measure* and is appropriate when it makes sense to attribute an equal weight or likelihood to each point of the sample space.

4.3. Example. In Example 2.1, we had the sample space

$$\Omega = \{(H, T), (T, H), (T, T), (H, H)\},$$

with the points of Ω describing the possible results of an experiment involving two successive tosses of a coin. The outcome (H, T) describes the outcome 'obtaining a head on the first toss and a tail on the second toss.' Let us form the power set 2^Ω of this sample space, that is, the set of all subsets of Ω. It is clear that 2^Ω is a field of sets of Ω. In fact, the power set of any arbitrary set X is a field of sets on X. (If this is not obvious, the student should examine again the definition of the power set 2^X of a set X, and the definition of a field of sets[1].)

The Probability Distribution

An interesting way of constructing a probability measure is provided by the following device. Assuming that the coin is fair in Example 2.1, we begin by defining a function p on Ω by: $p(x) = 1/4$, for all $x \in \Omega$. Next, we use p to obtain a probability measure $\mathbb{P}$, as a function on $\mathcal{F} = 2^\Omega$, using the definition

$$\mathbb{P}(A) = \sum_{x \in A} p(x), \quad \text{for all } A \subseteq \Omega. \tag{4.5}$$

4.4. Remark. A convenient notation is introduced informally here. If f is a nonnegative real valued function and A is a finite subset of the domain of f, then we write $\sum_{x \in A} f(x)$ for the sum of all the numbers $f(x)$, such that $x \in A$. This notation is justified because the order of the elements in the sum has no impact on the result. This convention is also appropriate when the set A is countable, and the right-hand side of (4.5) denotes the sum of a series.[2] We also

---

[1] We could even say that the power set of any arbitrary set is a σ-field. Why is that?

[2] As any calculus course would teach, this convention would not be correct for series without the specification that f is nonnegative.

set $\sum_{x \in \emptyset} f(x) = 0$. A consequence of (4.5) is, for example,

$$\mathbb{P}\{(H,T),(T,H)\} = p(H,T) + p(T,H) = (1/4) + (1/4) = 1/2,$$

that is, the probability of obtaining exactly one head in two tosses is equal to $1/2$. It is easy to see that $(\Omega, 2^{\Omega}, \mathbb{P})$ satisfies each of the three defining conditions of a finitely additive probability space. (The student should verify this.)

A simple method was used in this example to define a probability measure. We first assigned a number $p(x)$ to each outcome x in the sample space Ω in such a way that

$$p(x) \geq 0, \qquad \text{for all } x \in \Omega, \tag{4.6}$$

$$\sum_{x \in \Omega} p(x) = 1. \tag{4.7}$$

Next, we used the function p to define the probability measure $\mathbb{P}$ for all events A in the field of sets by setting $\mathbb{P}(A) = \sum_{x \in A} p(x)$.

This technique can always be used to define a probability measure, provided that the sample space Ω is a finite or countable set. Note that this technique generalizes the counting measure introduced in 4.2 in which each point x of the sample space Ω was assigned the same weight $\frac{1}{|\Omega|}$. Here, the weight assigned to the point x is $p(x)$. Let us formalize the construction of the probability measure used in these examples.

4.5. Definition. Let Ω be a finite or countable sample space. Then p is a *probability distribution on Ω* iff p is a real valued function on Ω satisfying (4.6) and (4.7).

Generalizing these examples, we have the following result:

4.6. Theorem. *Let $\mathcal{F}$ be a field of sets on a finite or countable sample space Ω, and let p be a probability distribution on Ω. Define a function $\mathbb{P}$ on $\mathcal{F}$ by*

$$\mathbb{P}(A) = \sum_{x \in A} p(x), \tag{4.8}$$

for all $A \in \mathcal{F}$. Then $(\Omega, \mathcal{F}, \mathbb{P})$ is a probability space.

PROOF. Axioms [K1] and [K2] of Def. 4.1 are immediate. Indeed, by (4.8) and the definition of a probability distribution, we have $\mathbb{P}(\Omega) = \sum_{x \in \Omega} p(x) = 1$, and it is clear that the function $\mathbb{P}$ is nonnegative.

Let $(A_i)_{i \in I}$ be a sequence of pairwise incompatible events, with the index set I being finite or countable, and each of the events A_i being finite or countable. Successively,

$$\sum_{i \in I} \mathbb{P}(A_i) = \sum_{i \in I} \sum_{x \in A_i} p(x) \qquad \text{[by definition of } \mathbb{P}(A_i) \text{ in (4.8)]}$$

$$= \sum_{x \in \cup_{i \in I} A_i} p(x)$$

$$= \mathbb{P}(\cup_{i \in I} A_i) \qquad \text{[by definition of } \mathbb{P}(\cup_{i \in I} A_i)\text{]}.$$

$\square$

A probability measure $\mathbb{P}$ constructed from a probability distribution p by the method of this theorem is said to be *induced* by p.

4.7. Example. (THE GAME SHOW CONTINUED.) The set of outcomes obtained for the game show was given in Table 2.1. For convenience, we reproduce the list of outcomes here (ignore the numbers in parentheses for the moment).

Table 4.1: List of outcomes in the game show

$\underline{A}^* B^\dagger C$	$(\frac{1}{18})$	$\underline{A} \, B^* C^\dagger$	$(\frac{1}{18})$	$\underline{A} \, B^\dagger C^*$	$(\frac{1}{18})$
$\underline{A}^* B \, C^\dagger$	$(\frac{1}{18})$	$A^\dagger \underline{B}^* C$	$(\frac{1}{9})$	$A^\dagger \underline{B} \, C^*$	$(\frac{1}{18})$
$A^* \underline{B} \, C^\dagger$	$(\frac{1}{9})$	$A\underline{B}^* C^\dagger$	$(\frac{1}{9})$	$A^\dagger B \, \underline{C}^*$	$(\frac{1}{9})$
$A^* B^\dagger \underline{C}$	$(\frac{1}{9})$	$A^\dagger B^* \underline{C}$	$(\frac{1}{18})$	$AB^\dagger \underline{C}^*$	$(\frac{1}{9})$

A reasonable probability distribution can be obtained from the following considerations. The 12 outcomes in this table fall into two classes:

1. The contestant chooses a wrong door (typical outcome: $A^* \underline{B} C^\dagger$). Six outcomes fall into that class.

2. The correct door is chosen (typical outcome: $\underline{A}^* B^\dagger C$). There are also six such outcomes.

It seems inescapable that all the outcomes within each class should be assigned the same probability. For instance, there would be no justification for deciding a priori that $A^*\underline{B}C^\dagger$ should be more probable than $\underline{A}B^*C^\dagger$. In Class 1, each outcome is specified by two features: (i) the correct door; (ii) the door selected by the contestant. Indeed, the host has no choice but to open the remaining door. The situation in Class 2 is similar except that the correct door coincides with the door selected by the contestant. In this case, the host has the choice between the two remaining doors. In Class 2, we have thus three cases, corresponding to the three possible correct doors. Each of these three cases splits then into two subcases, depending on the door opened by the host. It seems sensible that each of the three cases of Class 2 should have the same probability as that of any of the six cases of Class 2, that is: $1/9$. Thus, each of the six cases of Class 1 should have a probability of $1/9$, and each of the six cases of Class 2 should have a probability of $1/18$. All these probabilities are those given in parentheses in Table 4.1. Together, they define the probability distribution p in this situation. Using this probability distribution, we can then define a probability measure $\mathbb{P}$ specifying, for example, the probability of winning in the case of switching by the equation

$$\mathbb{P}\{\underline{A}B^*C^\dagger, \underline{A}B^\dagger C^*, A^\dagger \underline{B}C^*, A^*\underline{B}C^\dagger, A^*B^\dagger\underline{C}, A^\dagger B^*\underline{C}\}$$
$$= p(\underline{A}B^*C^\dagger) + p(\underline{A}B^\dagger C^*) + p(A^\dagger \underline{B}C^*) + p(A^*\underline{B}C^\dagger)$$
$$+ p(A^*B^\dagger\underline{C}) + p(A^\dagger B^*\underline{C})$$
$$= \frac{1}{9} + \frac{1}{9} + \frac{1}{9} + \frac{1}{9} + \frac{1}{9} + \frac{1}{9} = \frac{2}{3}.$$

The event 'Winning in the case of staying' contains the six outcomes with probability $\frac{1}{18}$. Accordingly, the probability of that event is $\frac{1}{3}$. The switching strategy is thus correct.

4.8. Remark. The material that was just covered correctly suggests that if only finite or countable sample spaces were considered, the notions of a field of events and of a probability measure could be dispensed with entirely because the probability of any subset of the sample space could be computed from the probability distribution.

With uncountable sample spaces,[3] the notion of a probability distribution is useless. Two examples of uncountable sample spaces were encountered earlier: the reaction time experiment of Example 2.5, in which the appropriate sample space was the set of all

---

[3] For instance, in the guise of the set of real numbers or the set of all sequences in a finite set (both of which play an important role in the applications).

positive real numbers, and the T-maze experiment of Example 2.3, in which the sample space was the set of all sequences in the set $\{L, R\}$ (where L and R represent, respectively, a left and right turn in the maze).

However, it turns out that in many important special cases of an uncountable sample space, there is a notion conceptually related to that of a probability distribution, which plays a similar role in permitting the construction of a unique probability measure. This concept is called a 'probability density function' and is covered in Chapter 14.

Problems

1. Verify that the counting measure of Example 4.2 is a probability measure.

2. Let $\Omega = \{a, b, c\}$ be a sample space, and suppose that the collection of all events is the power set of Ω. You know part of a probability measure $\mathbb{P}$, and part of a probability distribution p. Specifically, $\mathbb{P}\{a, b\} = .7$ and $p(b) = .1$. Provide the missing values of $\mathbb{P}$ and p in such a way that p is consistent with $\mathbb{P}$ in the sense of the equation $\mathbb{P}(A) = \sum_{x \in A} p(x)$. Is there only one way to do this?

3. Suppose that there are 5 hats in a trunk: 2 white hats and 3 black ones. Each of three brothers is randomly picking a hat in the trunk. (a) Construct an appropriate sample space for this experiment. How many points does it have? (b) What is the probability that the three brothers have a black hat? (c) That two of the brothers have a white hat?

4. A fair die is rolled twice. What are the probabilities of the following events: (a) the second number is twice the first? (b) the second number is not greater than the first? (c) at least one number is greater than 3?

5. Construct an example in which $1 + \mathbb{P}(A \cap B) \geq \mathbb{P}(A) + \mathbb{P}(B)$.

6. Is it possible that $\mathbb{P}(A \cap B) < \mathbb{P}(A)\mathbb{P}(B)$? (When $\mathbb{P}(A \cap B) = \mathbb{P}(A)\mathbb{P}(B)$ the events A and B are said to be 'independent'; see 10.1.)

7. Define a probability distribution p on $\mathbb{N}$ such that $p(n) > 0$ for all $n \in \mathbb{N}$.

8. Compute the probability that a triangle may be formed from the three resulting fragments when a thin, straight rod of unit length in broken in two places at random. Assume that all break points are equally likely. (Cf. Problem 5 in Chapter 2; Beckmann, 1967.)

9. Consider the set of all rankings with possible ties on the set $\{a_1, a_2, a_3\}$ already encountered in Problem 6 of Chapter 3. Suppose that all rankings are equally likely. Compute the probabilities of the following three events: (a) a_2 is ranked above a_1; (b) there is no tie; (c) there is a tie between a_2 and a_3.

10. Suppose that p_1 is a probability distribution on Ω_1 and that p_2 is a probability distribution on Ω_2. Define a function p on $\Omega_1 \times \Omega_2$ by the equation $p(x_1, x_2) = p_1(x_1)p_2(x_2)$, which is assumed to hold for all (x_1, x_2) in $\Omega_1 \times \Omega_2$. Verify that p is a probability distribution on $\Omega_1 \times \Omega_2$.

11. In the variant of the game show presented in Problem 2 of Chapter 2, is the contestant still advised to switch? What about variants in which the number of doors is five or more?

12. The U.S. Senate has 91 male members and 9 female members. Suppose that a Senate committee position must be filled, and that those of the same sex have equal probability of being chosen. Suppose also that each woman is twice as likely to be chosen as each man. Evaluate the probabilities of the following events: (a) A woman (any woman) is chosen; (b) One of the two male senators from Indiana is chosen; (c) Neither of the two female senators from California is chosen.

Chapter 5

Basic Rules of Probability Calculus

This chapter is devoted to a presentation of a number of consequences of Definition 4.1 of a finitely additive probability space $(\Omega, \mathcal{F}, \mathbb{P})$. These consequences are listed in the next theorem. NOTE THAT WE DO NOT ASSUME THAT Ω OR $\mathcal{F}$ ARE FINITE. This means that these results apply to the case of a countably additive measure on a σ-field of events.

Main Results

5.1. Theorem. *For any events A, B, and C:*

(i) $\mathbb{P}(A) + \mathbb{P}(\overline{A}) = 1$;

(ii) $\mathbb{P}(\emptyset) = 0$;

(iii) *if $A_1, A_2, \ldots, A_n$ are pairwise incompatible events, then*
$$\mathbb{P}(\cup_{i=1}^{n} A_i) = \sum_{i=1}^{n} \mathbb{P}(A_i);$$

(iv) *if $A \subseteq B$, then $\mathbb{P}(A) \leq \mathbb{P}(B)$;*

(v) $\mathbb{P}(A \cup B) = \mathbb{P}(A) + \mathbb{P}(B) - \mathbb{P}(A \cap B)$;

(vi) $\mathbb{P}(A \cup B \cup C) = \mathbb{P}(A) + \mathbb{P}(B) + \mathbb{P}(C) - \mathbb{P}(A \cap B)$
$$- \mathbb{P}(A \cap C) - \mathbb{P}(B \cap C) + \mathbb{P}(A \cap B \cap C);$$

(vii) $\mathbb{P}(A \cap \overline{B}) = \mathbb{P}(A) - \mathbb{P}(A \cap B) = \mathbb{P}(A \setminus B)$.

5.2. Convention. We recall that when two events A, B are pairwise incompatible, we often write $A + B$ to mean $A \cup B$. Similarly, for a collection $A_1, A_2, \ldots, A_n$ of pairwise incompatible events, we may write

$$\sum_{i=1}^{n} A_i = \bigcup_{i=1}^{n} A_i.$$

This convention will be used freely in the proof of Theorem 5.1. We shall also refer to the three axioms defining a finitely additive probability space in Definition 4.1 by their labels [K1], [K2], and [K3], which the student is advised to memorize.

PROOF OF THEOREM 5.1. (i) From set theory, by the definition of $\overline{A}$, we have $A \cap \overline{A} = \emptyset$ and $A + \overline{A} = \Omega$. Using Axioms [K3] and [K1], we have

$$\mathbb{P}(A) + \mathbb{P}(\overline{A}) = \mathbb{P}(A + \overline{A}) = \mathbb{P}(\Omega) = 1.$$

(ii) We have successively:

$$1 = \mathbb{P}(\Omega) = \mathbb{P}(\Omega + \emptyset) = \mathbb{P}(\Omega) + \mathbb{P}(\emptyset) = 1 + \mathbb{P}(\emptyset).$$

This yields $1 + \mathbb{P}(\emptyset) = 1$, and thus $\mathbb{P}(\emptyset) = 0$.

(iii) We use induction on the number n of events. For $n = 1$, we have trivially $\mathbb{P}(\cup_{i=1}^{1} A_i) = \mathbb{P}(A_1) = \sum_{i=1}^{1} \mathbb{P}(A_i)$. Suppose that (iii) holds for any $n = m \geq 1$, and consider $n = m + 1$. Successively,

$$
\begin{aligned}
\mathbb{P}(\cup_{i=1}^{m+1} A_i) &= \mathbb{P}[(\cup_{i=1}^{m} A_i) + A_{m+1}] && \text{(set theory)} \\
&= \mathbb{P}(\cup_{i=1}^{m} A_i) + \mathbb{P}(A_{m+1}) && \text{(by Axiom [K3])} \\
&= \sum_{i=1}^{m} \mathbb{P}(A_i) + \mathbb{P}(A_{m+1}) && \text{(by the induction hypothesis)} \\
&= \sum_{i=1}^{m+1} \mathbb{P}(A_i), && \text{(rearranging)}
\end{aligned}
$$

establishing (iii).

(iv) The hypothesis $A \subseteq B$ implies $B = A + (B \setminus A)$. Hence, from Axiom [K3], $\mathbb{P}(B) = \mathbb{P}[A + (B \setminus A)] = \mathbb{P}(A) + \mathbb{P}(B \setminus A)$, with $\mathbb{P}(B \setminus A) \geq 0$, yielding the result.

(v) For any two events $A, B \in \mathcal{F}$, we have in general,

$$A \cup B = (A \setminus B) + (B \setminus A) + (A \cap B). \tag{5.1}$$

We also observe that

$$A = (A \setminus B) + (A \cap B), \tag{5.2}$$
$$B = (B \setminus A) + (A \cap B). \tag{5.3}$$

From (5.1), (5.2), and (5.3), we derive, using [K3]:

$$\mathbb{P}(A \cup B) = \mathbb{P}(A \setminus B) + \mathbb{P}(B \setminus A) + \mathbb{P}(A \cap B), \tag{5.4}$$
$$\mathbb{P}(A \setminus B) = \mathbb{P}(A) - \mathbb{P}(A \cap B), \tag{5.5}$$
$$\mathbb{P}(B \setminus A) = \mathbb{P}(B) - \mathbb{P}(A \cap B). \tag{5.6}$$

Adding (5.4), (5.5), and (5.6) and cancelling appropriately gives (v).

(vi) We only give a sketch of this proof, which relies on repeated applications of (v) that was just obtained. We simply state the successive computations without justification. Successively,

$$
\begin{aligned}
\mathbb{P}(A \cup B \cup C) \\
&= \mathbb{P}((A \cup B) \cup C) \\
&= \mathbb{P}(A \cup B) + \mathbb{P}(C) - \mathbb{P}((A \cup B) \cap C) \\
&= \mathbb{P}(A) + \mathbb{P}(B) - \mathbb{P}(A \cap B) + \mathbb{P}(C) - \mathbb{P}((A \cap C) \cup (B \cap C)) \\
&= \mathbb{P}(A) + \mathbb{P}(B) - \mathbb{P}(A \cap B) + \mathbb{P}(C) \\
&\quad - (\mathbb{P}(A \cap C) + \mathbb{P}(B \cap C) - \mathbb{P}((A \cap C) \cap (B \cap C))) \\
&= \mathbb{P}(A) + \mathbb{P}(B) + \mathbb{P}(C) - \mathbb{P}(A \cap B) - \mathbb{P}(A \cap C) \\
&\quad - \mathbb{P}(B \cap C) + \mathbb{P}(A \cap B \cap C).
\end{aligned}
$$

(vii) Set theory yields $A = (A \cap \overline{B}) + (A \cap B) = (A \setminus B) + (A \cap B)$. The result of (vii) follows from Axiom [K3] of a finitely additive probability space. $\square$

Poincaré's Identity

Formulas (v) and (vi) in Theorem 5.1 can be generalized under the name of *Poincaré's Identity* to any arbitrary number of events. In the theorem below, we write $S(k, n)$ for the collection of subsets of the set $\{1, \dots, n\}$ containing exactly k elements, with $1 \le k \le n$.

5.3. Theorem. (POINCARÉ'S IDENTITY.)   *For any finite collection A_1, A_2, ... , A_n of events, we have*

$$\mathbb{P}(\cup_{i=1}^n A_i) = \sum_{i=0}^{n-1}(-1)^i \sum_{J \in S(i+1,n)} \mathbb{P}(\cap_{k \in J} A_k). \qquad (5.7)$$

Note that $(-1)^i$ is equal to 1 or -1, depending on whether i is even or odd, respectively.

SKETCH OF PROOF. The case $n = 1$ of this theorem is trivial. Theorem 5.1 (v) and (vi) correspond to the cases $n = 2$ and $n = 3$ of Theorem 5.3. The rest of the proof is by induction. Suppose that (5.7) holds for $n = m$, and consider $n = m + 1$. By Theorem 5.1 (v), we get

$$\mathbb{P}(\cup_{i=1}^{m+1} A_i) = \mathbb{P}(\cup_{i=1}^m A_i) + \mathbb{P}(A_{m+1}) - \mathbb{P}\left((\cup_{i=1}^m A_i) \cap A_{m+1}\right)$$
$$= \mathbb{P}(\cup_{i=1}^m A_i) + \mathbb{P}(A_{m+1}) - \mathbb{P}\left(\cup_{i=1}^m (A_i \cap A_{m+1})\right).$$

Each of the two unions in the last equation is taken over m sets. By the induction hypothesis, we can apply Poincaré's identity for the particular value m to both $\mathbb{P}(\cup_{i=1}^m A_i)$ and $\mathbb{P}\left(\cup_{i=1}^m (A_i \cap A_{m+1})\right)$. We obtain

$$\mathbb{P}(\cup_{i=1}^{m+1} A_i) = \overbrace{\sum_{i=0}^{m-1}(-1)^i \sum_{J \in S(i+1,m)} \mathbb{P}(\cap_{k \in J} A_k) + \mathbb{P}(A_{m+1})}^{\mathbf{C}}$$

$$\underbrace{- \sum_{i=0}^{m-1}(-1)^i \sum_{J \in S(i+1,m)} \mathbb{P}(\cap_{k \in J}(A_k \cap A_{m+1}))}_{\mathbf{D}}. \quad (5.8)$$

To complete our induction argument, we have to show that

$$\sum_{i=0}^{m}(-1)^i \sum_{J \in S(i+1,m+1)} \mathbb{P}(\cap_{k \in J} A_k) = \mathbf{C} + \mathbb{P}(A_{m+1}) - \mathbf{D}.$$

All the terms in Subformula $\mathbf{C}$ appear also, with the same sign, in the left hand side of the above equation. Missing from $\mathbf{C}$ are the probabilities of intersection of A_{m+1} with all the intersections of the subcollections of $\{A_1, \ldots, A_m\}$. The student should verify that all these missing terms (except $\mathbb{P}(A_{m+1})$) appear in Subformula **D** with the correct sign (Problem 3). Thus, grouping terms appropriately in the right-hand side of (5.8) gives the result of the theorem.   □

Problems

1. Let A and B be disjoint events, with $0 < \mathbb{P}(A) + \mathbb{P}(B) < 1$. Express in terms of $x = \mathbb{P}(A)$ and $y = \mathbb{P}(B)$ the probabilities of the events:

   (a) A occurs but B does not occur;

   (b) either A does not occur or B does not occur;

   (c) either A occurs or B does not occur;

   (d) neither the event A nor the event B occurs;

   (e) either A but not B occurs, or B but not A occurs;

   (f) A but not B occurs; B but not A fails to occur.

2. Justify all the steps in the sketch of the proof of Theorem 5.1 (vi).

3. Verify carefully and fill in the details of the induction argument sketched in the proof of Theorem 5.3.

4. In a certain sports league there are four teams: a_1, a_2, a_3, and a_4. It is estimated that the probability that a_1 finishes first is $\frac{1}{2}$, the probability that a_2 finishes last is $\frac{1}{2}$, the probability that a_3 finishes above a_4 is $\frac{1}{4}$, and the probability that a_1 finishes first or a_2 finishes last (or both) is $\frac{3}{4}$. What is the maximum possible probability for the event 'either a_1 finishes first and a_2 finishes last, or a_3 is ranked above a_4, or both'?

5. In a certain city, 53% of the adults are female and 15% are unemployed males. (a) What is the probability that an adult chosen at random in this city is an employed male? (b) If the overall unemployment rate is 22%, what is the probability that an adult is an employed female? (c) What is the probability that an adult is employed or female (or both)?

6. Recall the approval voting scheme of Problem 6, Chapter 2, in which a voter may approve any subset of the candidates. In one such election, candidate c_1 is approved of by 53% of the voters, candidate c_2 by 60% of the voters, and candidate c_3 by 77% of the voters. In addition, c_1 and c_2 are together approved by 45% of the voters. Based on this information, what can you say about the percentages for the approval of both c_2 and c_3, the approval of both c_1 and c_3, and the approval of all of c_1, c_2, and c_3?

7. 87.5% of the residents of a small town in California were born in the United States. Of the 12.5% who are immigrants, 99% speak Spanish fluently, while 13% of all residents speak Spanish fluently. Suppose that a resident of the town is randomly chosen. Rank the following from least likely to most likely:
   A: speaks Spanish fluently
   B: was born in the United States and speaks Spanish fluently
   C: is an immigrant who speaks Spanish fluently
   D: was not born in the United States.

8. In five tosses of a fair coin, what is the probability that the first head does not appear until the third toss or there are at least three straight heads (or both)?

9. The *symmetric difference* between two sets A and B is defined by the equation $A \Delta B = A \cap \overline{B} + B \cap \overline{A}$. Prove that, for events A and B, we have $\mathbb{P}(A \Delta B) = \mathbb{P}(A) + \mathbb{P}(B) - 2\mathbb{P}(A \cap B)$. (Do not rely on Venn diagrams.)

10. Use induction to prove *Boole's Inequality*: $\mathbb{P}(\cup_{i=1}^{n} A_i) \leq \sum_{i=1}^{n} \mathbb{P}(A_i)$, for any arbitrary events $A_1, \ldots, A_n$.

11. Using Poincaré's Identity (Theorem 5.3) as a guide, find a simple expression for $|\cup_{i=1}^{n} A_i|$, where $A_1, A_2, \ldots, A_n$ are subsets of a finite set X.

*12. Generalizing Eq. (2.4), find an expression in the style of Poincaré's Identity for indicator functions.

13. Suppose that four people each toss a fair coin ten times. Discuss the appropriateness of the calculation $(4)(\frac{1}{2^{10}})$ for estimating the probability that at least one of the four people obtains ten consecutive heads.

Chapter 6

Sampling

When a sample space Ω is finite and the counting measure is used (cf. 4.2), the probability of an event A is defined as the ratio

$$\mathbb{P}(A) = \frac{|A|}{|\Omega|}.$$

This probability is thus computed by counting the number of points in the event A and dividing it by the number of points in the sample space. Such a measure is appropriate in those situations when all the points of the sample space are regarded as being equally likely. Examples are easily manufactured. We describe two of them below.

6.1. Example. (THE CARD GAME.) In a card game using the standard 52-card deck, each of four players receives 13 cards. Assuming that the cards have been thoroughly shuffled, what is the probability that one of the players receives all the clubs? We take as our sample space the collection of all partitions of the deck into four subsets of 13 (for 'partition,' see 1.7). Idealizing this situation somewhat, we suppose that all those partitions are equally likely.[1] In principle, the required probability can be obtained by counting the number of partitions in which one of the players receives all the clubs, and dividing it by the number of points in the sample space, that is, the total number of partitions of the set $\{1, \ldots, 52\}$ into four classes, each containing 13 points.

---

[1] In practice, even after careful shuffling, the assumption that all the partitions are equally likely will not hold.

6.2. Example. (THE BIRTHDAY PROBLEM.) Twenty people are gathered in a room, and it is discovered that two of them have the same birthday. What is the probability of this event? We take as our sample space the set of all 20-tuples of natural numbers $(x_1, \dots, x_{20})$, with $1 \le x_i \le 365$ for $i = 1, \dots, 20$. We suppose that all birthdays are equiprobable.[2] Specifically, the probability that a randomly sampled individual has his or her birthday on any particular day of the year is $\frac{1}{365}$, and all the points $(x_1, \dots, x_{20})$ of the sample space have the same probability. The required probability is obtained by counting all the points $(x_1, \dots, x_{20})$ of the sample space having two (or more?) identical terms ($x_i = x_j$ for some $i \ne j$) and dividing it by the total number of points in the sample space. This problem and the preceding one are solved in Chapter 7.

The next example shows that counting techniques may be useful even in cases in which the points of the sample space are not equiprobable.

6.3. Example. (BALLS IN AN URN.) Eight balls are sampled from an urn containing 10% black balls and 90 % white ones, replacing the ball in the urn after each selection. (This is called 'sampling with replacement'; see 6.7.) What is the probability of obtaining exactly 5 white balls? The natural sample space to use is the set of all 8-tuples $(x_1, \dots, x_8)$, with $x_i = 0, 1$ (for $1 \le i \le 8$) representing black and white, respectively. The points of this sample space are not equiprobable. (For instance, the outcome $(0, 0, \dots, 0)$ is much less likely than the outcome $(1, 1, \dots, 1)$.) However, any outcome represented by an 8-tuple with five 1s and three 0s is as likely as any other such outcome. For instance $(1, 1, 1, 0, 0, 1, 0, 1)$ is as likely as $(1, 1, 1, 1, 1, 0, 0, 0)$. Let θ be the probability of one such outcome, say $(1, 1, 1, 0, 0, 1, 0, 1)$. The probability of obtaining exactly 5 white balls can be obtained by counting all such outcomes in the sample space. If that number is K, then the probability of obtaining exactly 5 white balls in drawing 8 balls is equal to $K\theta$. A more complete discussion of this example is given in Chapter 8, in the context of the so-called 'binomial distribution.'

In each of these problems, the calculation of a probability relies on counting the number of points in one or more events. However,

---

[2] In practice, this assumption is not valid. Consequently, the computed probability will be underestimated.

how to perform such counting accurately in practice is not immediately obvious. This chapter and the next one provide a number of counting techniques to solve this kind of problem. Some of these tools are organized under the headings of two fundamental kinds of sampling that arise in probability and statistics: 'sampling with replacement' and 'sampling without replacement.' These tools will equip the student with a collection of formulas than can be used to facilitate or organize the counting. However, it must be understood that the application of such formulas is never automatic. The choice of the formulas which are adequate to a particular problem always relies on a correct conceptual analysis of the problem.

Sampling with Replacement and with Ordering

We begin with a preparatory concept: the number of points in a Cartesian product (see 1.5).

6.4. Example. A social scientist interviews individuals in a mixed population. The individuals are classified according to their gender (2), ethnic group (5), social class (5), salary range (6) and occupation (8). A priori, the total number of possible classes in which an individual may be assigned is equal to $2 \times 5 \times 5 \times 6 \times 8 = 2400$.

The proof of the following basic result is left to the reader.

6.5. Theorem. *Let* A_1, A_2, ... , A_n *be* n *finite sets containing* m_1, m_2, ... , $m_n > 0$ *points, respectively. Then*

$$|A_1 \times A_2 \times \cdots \times A_n| = m_1 \cdot m_2 \cdot \ldots \cdot m_n.$$

The concept of 'sampling with replacement and with ordering' is closely related to this theorem.

6.6. Example. An biologist selects a sample of fish from a pool, replacing each animal in the pool after marking it in some fashion. If the total population of fish in the pool contains 100 individuals, how many different ordered samples of size 8 are there? By 'ordered sample,' I mean that the experimenter keeps track of the order in which the animals are selected. There are 100 possibilities for the choice of the first animal. Since the animal is replaced in the population after marking, there are also 100 possibilities for the second animal, and so on. Thus, there are 100^8 possible samples overall.

6.7. Definition. In statistics, the term *ordered sample of size n with replacement* from a set X is used to describe a point of the Cartesian product X^n, that is, an n-tuple $(x_1, \ldots, x_n)$ of elements of the set X.

6.8. Theorem. *In a set containing $n > 0$ elements, there are exactly n^m ordered samples of size $m \geq 1$, with replacement.*

PROOF. This is a special case of Theorem 6.5.   □

Sampling without Replacement

We still deal with ordered samples, that is, we keep track of the order in which the elements are sampled. Here, however, we do not replace the element in the set after each drawing. We first examine the possible ways of ordering a set.

6.9. Example. The elements of the set $\{a, b, c, d\}$ can be ordered in 24 ways:

abcd	*abdc*	*acbd*	*acdb*	*adbc*	*adcb*
bacd	*badc*	*bcad*	*bcda*	*bdac*	*bdca*
cabd	*cadb*	*cbad*	*cbda*	*cdab*	*cdba*
dabc	*dacb*	*dbac*	*dbca*	*dcab*	*dcba*.

(By comparison, the number of ordered samples of size 4, with replacement, is $256 = 4^4$.) To formulate a general result, we introduce a compact notation.

6.10. Definition. The *factorial* of a natural number n is the natural number

$$n! = n(n-1)(n-2)\ldots 2 \cdot 1.$$

It is also convenient to define

$$0! = 1.$$

6.11. Theorem. *A set containing $n \geq 1$ elements can be ordered in $n!$ ways.*

PROOF. We have n possibilities for the choice of the first element in the order. Once the first element has been chosen, $n - 1$ possibilities remain for the choice of the second element. Hence, we have $n(n - 1)$ possibilities for the choice of the first two elements. Similarly, we have $n(n - 1)(n - 2)$ possibilities for the choice of the first three elements, etc. $\qquad\square$

(The reader may not be impressed by this intuitive argument. Problem 1 at the end of the chapter requires the student to provide a proof by induction.) The concept of 'sampling without replacement and with ordering' generalizes the situation in Theorem 6.11.

6.12. Example. An experimenter selects 4 rats (without replacement) from a population of 10 rats. How many different ordered samples are possible? There are 10 rats to choose from at the beginning of the selection. After choosing the first rat, nine rats remain. There are thus $10 \cdot 9$ possibilities for the choice of the first two rats. Pursuing the selection process, we obtain $10 \cdot 9 \cdot 8 \cdot 7 = 5040$ different samples of size 4, without replacement, in a set of size 10.

6.13. Definition. The term *ordered sample of size n without replacement* from a set X is used in statistics to denote any n-tuple $(x_1, \ldots, x_n)$ of DISTINCT elements of X.

The next theorem generalizes Example 6.12.

6.14. Theorem. *The number of ordered samples of size m without replacement, in a set of $n \geq m > 1$ elements, is equal to*

$$(n)_m = n(n - 1) \cdot \ldots \cdot (n - m + 1) = \frac{n!}{(n - m)!}$$

The first equality defines the notation $(n)_m$. We have thus $(n)_n = n!$.

Stirling's Formula

In practice, computation involving factorials can be facilitated by approximation formulas. The best known is due to Stirling (1730), who proved the following result:

6.15. Theorem. *For any positive integer n, we have*

$$n! \approx (2\pi)^{\frac{1}{2}} n^{n + \frac{1}{2}} e^{-n}. \tag{6.1}$$

In this formula, $\approx$ means that the ratio in the two sides of (6.1) tends to 1 as $n \to \infty$. A proof of this approximation is contained in Feller (1957).

A more accurate approximation has been given by Robbins (1955), who proved the following double inequality:

$$(2\pi)^{\frac{1}{2}} n^{n+\frac{1}{2}} e^{\frac{-n+1}{12n+1}} < n! < (2\pi)^{\frac{1}{2}} n^{n+\frac{1}{2}} e^{\frac{-n+1}{12n}}. \qquad (6.2)$$

Problems

1. Provide an inductive proof of Theorem 6.11.

2. Check the approximation of $n!$ provided by (6.1) for $n = 11$. Compare the result with the bounds provided by (6.2).

3. How many ordered samples of size five, without replacement, are there in a set containing 30 elements? Solve this problem in two ways: directly and using Stirling's formula.

4. (a) Justify the remark in footnote 2 on the second page of this chapter. In particular, verify that the fact that birthdays are more common in the months of September and October than in other months gives an *under-estimation* of the shared birthday probability. (b) Another simplification is made in the birthday problem, viz., that there are only 365 possible birthdays. How does this affect the resulting probability estimate?

5. (a) Suppose that an ordered sample, with replacement, of size three is selected from a set of five elements. What is the probability that the sample obtained is an ordered sample without replacement (that is, that all the terms of the 3-tuple are different)? (b) What is the probability that an ordered sample, with replacement, of size m selected from a set of n elements ($n \geq m > 1$) is an ordered sample without replacement?

6. How many words can be created from the word SAMPLE? A created word does not have to be an actual English word, but it may contain at most as many instances of a letter as there are in the original word (for example, 'ama' is not acceptable, whereas 'pma' is).

7. In how many ways can n people be arranged in a line? In how many ways can they be arranged in a circle? One or both of these questions may be ambiguous—you should analyze the different interpretations.

8. The standard car license plate in a certain state has seven characters. The first character is one of the digits 1, 2, 3, or 4; the next three characters are letters (repetitions allowed); and the final three characters are digits (0, 1, ... , 9; repetitions are allowed). (a) How many license plates are

possible? (b) How many have no repeated characters? (c) How many have a vowel $(A, E, I, O,$ or $U)$ as the first letter or the sequence 222 as the final three digits? (d) Assuming that each license plate is equally likely, what is the probability that a license plate chosen at random ends with your favorite digit?

9. A magician is guessing the names of four cards labeled 1 to 4 placed faced-down on a table. Assuming that he is a faker, what is the probability that he guesses at least one card right? Note that the magician guesses all four cards before turning them over. (Hint: Just analyze all the possibilities carefully.)

10. Five boys and five girls are getting together for a party. How many couples can be formed? Suppose that one of the boys has two sisters among the five girls, and would not accept either of them as a partner. How many couples can be formed then?

Chapter 7

Counting Subsets

In spaces equipped with a counting measure (cf. 4.2), many probability problems encounter the following type of question: How many subsets of size m are there in a set of size $n \geq m$? This question may seem difficult, but is actually straightforward if we use the results established in Chapter 6.

The Binomial Coefficient

Let us denote by $\binom{n}{m}$ the unknown number of subsets of size $m \geq 0$ in a set containing n elements. Notice that each of these $\binom{n}{m}$ subsets can be ordered in $m!$ different ways (Theorem 6.11). Consequently, the product $\binom{n}{m} \cdot m!$ is the number of ordered samples of size m, without replacement, in a set of size n. Using Theorem 6.14, we obtain

$$\frac{n!}{(n-m)!} = \binom{n}{m} \cdot m!.$$

Solving for $\binom{n}{m}$ in the above equation yields the following result:

7.1. Theorem. *In a set of size $n \geq 0$, the number of subsets of size $m \geq 0$ is equal to*

$$\binom{n}{m} = \frac{n!}{m!(n-m)!}. \tag{7.1}$$

Notice in particular that

$$\binom{n}{0} = \binom{n}{n} = \frac{n!}{n!\,0!} = 1.$$

(There is exactly one subset containing zero elements, namely the empty set $\emptyset$, and one subset containing n elements, namely the set itself.)

7.2. Definition. The coefficient $\binom{n}{m}$ defined by (7.1) is called the *binomial coefficient*. Note that we also define

$$\binom{n}{m} = 0 \qquad \text{for } n < m \text{ or } m < 0. \tag{7.2}$$

The binomial coefficient plays an important role in probability and statistics. A noteworthy application is the so-called *Binomial Theorem*,[1] which is formulated below.

The Binomial Theorem

7.3. Theorem. *Let a and b be any two real numbers, and let n be a positive integer. Then*

$$(a + b)^n = \sum_{k=0}^{n} \binom{n}{k} a^k b^{n-k}. \tag{7.3}$$

PROOF. Developing the left hand side of (7.3) yields

$$\underbrace{(a + b)(a + b)\ldots(a + b)}_{n \text{ times}}$$

$$= C(0)a^n + \ldots + C(k)a^{n-k}b^k + \ldots + C(n)b^n, \tag{7.4}$$

for some coefficients $C(0), C(1), \ldots, C(n)$. Indeed, multiplying out the factors in the left hand side of (7.4) gives a sum of terms each of which must have the form $C(k)a^{n-k}b^k$ ($0 \leq k \leq n$). The coefficient $C(k)$ indicates the number of different ways the product $a^{n-k}b^k$ occurs in multiplying out $(a + b)^n$. It only remains to show that $C(k) = \binom{n}{k}$ for $k = 0, 1, \ldots, n$. Notice that $C(0)$ must be equal to 1 because the only way of forming a^n is to take a in each of the n factors $(a + b)$. Thus, $C(0) = \binom{n}{0}$. For similar reasons, there is a single term b^n yielding $C(n) = \binom{n}{n}$. We must have $C(1) = n = \binom{n}{1}$ because there are n different ways of forming the product $a^{n-1}b$:

---

[1]This result is often attributed to Isaac Newton, who is said to have proved it at the age of nineteen. However, Newton had been anticipated by Nicolo Tartaglia (1500-1557). This result was actually known even earlier by the Arab mathematicians (see Yaglom, 1988, footnote 12).

the single b in that expression may come from each of the n factors in $(a+b)^n$. In the language of this chapter, there are $\binom{n}{1} = n$ different ways of picking a subset of size one in a set containing n elements. This argument can be generalized. The coefficient $C(k)$ of the generic term $C(k)a^{n-k}b^k$ must be equal to $\binom{n}{k}$ because counting the number of ways the product $a^{n-k}b^k$ is formed in multiplying out $(a+b)^n$ corresponds to counting the number of subsets of size k in a set containing n elements (cf. Theorem 7.1). $\square$

We formulate a couple of identities for binomial coefficients.

7.4. Theorem. *For $0 \le m \le n$,*

$$\binom{n}{m} = \binom{n}{n-m} \tag{7.5}$$

$$\binom{n}{m} = \binom{n-1}{m-1} + \binom{n-1}{m} \tag{7.6}$$

$$2^n = \binom{n}{0} + \binom{n}{1} + \dots \binom{n}{n} = \sum_{k=0}^{n} \binom{n}{k}. \tag{7.7}$$

PROOF. Equation (7.5) follows immediately from Eq. (7.1). Equation (7.6) can also be established by computation, but the following conceptual proof is more interesting. According to Theorem 7.1, there are $\binom{n}{m}$ subsets of size m in a set S of size n. Let a be any element of S, which we keep fixed for this argument. The $\binom{n}{m}$ subsets of S can be partitioned into two classes: (1) those containing a, and there are $\binom{n-1}{m-1}$ of them (because they can be counted by removing a from S and counting the number of subsets of size $m-1$ in a set of $n-1$ elements); (2) those not containing a, and there are $\binom{n-1}{m}$ (because they can be counted by removing a from S and counting the number of subsets of size m in a set of $n-1$ elements). The total number of subsets of size m must be the sum of these two numbers. Equation (7.7) immediately results from the Binomial Theorem 7.1, by setting $a = b = 1$. $\square$

7.5. Remark. Another proof of (7.7) is obtained from remarking that the right hand side of that equation is summing the number of subsets of size $0, 1, \dots, n$. Since the total number of subsets in a set of n elements is equal to 2^n, the result follows.

The Multinomial Coefficient

In the last section, we learned that there were $\binom{n}{m}$ ways of splitting a set of n elements into two subsets containing $n - m$ and m elements, respectively (with $0 \le m \le n$). This result is begging to be generalized.

7.6. Problem. In how many ways can be split a set S of n elements into k disjoint subsets $S_1,\ldots,S_k$ containing $m_1,\ldots,m_k \ge 0$ elements, respectively, with $\sum_{i=1}^{k} m_i = n$? To avoid any misunderstanding, an example is given below. We take $S = \{a,b,c,d\}$, and we consider the collection of all triples of disjoint sets (S_1, S_2, S_3) such that $|S_1| = 2$ and $|S_2| = |S_3| = 1$ (thus, $m_1 = 2$ and $m_2 = m_3 = 1$). There are 12 such triples:

$$
\begin{array}{ll}
(\{a,b\},\{c\},\{d\}) & (\{a,b\},\{d\},\{c\}) \\
(\{a,c\},\{b\},\{d\}) & (\{a,c\},\{d\},\{b\}) \\
(\{a,d\},\{c\},\{b\}) & (\{a,d\},\{b\},\{c\}) \\
(\{b,c\},\{a\},\{d\}) & (\{b,c\},\{d\},\{a\}) \\
(\{b,d\},\{a\},\{c\}) & (\{b,d\},\{c\},\{a\}) \\
(\{c,d\},\{a\},\{b\}) & (\{c,d\},\{b\},\{a\})
\end{array}
\tag{7.8}
$$

Note that each of the two triples of sets on each line corresponds to the same partition. There are thus two differences between this counting problem and that of counting partitions. By definition, the classes of a partition are not empty, whereas here, because of the way the problem has been raised, some of the sets S_i may be empty. The other difference is that we are counting k-tuples of sets, and two or more k-tuples may correspond to the same partition.

To solve the general case, we proceed gradually. We first pick S_1, which contains m_1 elements. We know by Theorem 7.1 that there are $\binom{n}{m_1}$ different ways to pick such a subset. Now, the second subset S_2 of m_2 elements has to be picked from the $m_2 + m_3 + \ldots + m_k$ elements remaining after the removal of S_1. By the same argument, there are

$$
\binom{m_2 + m_3 + \ldots + m_k}{m_2}
$$

different ways of picking this second subset. Overall, there are

$$\binom{n}{m_1}\binom{m_2 + m_3 + \ldots + m_k}{m_2}$$

$$= \frac{n!}{(m_1)!(m_2 + m_3 + \ldots + m_k)!} \times \frac{(m_2 + m_3 + \ldots + m_k)!}{m_2!(m_3 + \ldots + m_k)!}$$

$$= \frac{n!}{m_1! m_2!(m_3 + \ldots + m_k)!}$$

different ways of choosing the first two subsets, containing m_1 and m_2 elements, respectively. Pursuing this way up to the last subset of m_k elements, the following result clearly obtains:

7.7. Theorem. *There are exactly*

$$\binom{n}{m_1 \quad m_2 \quad \ldots \quad m_k} = \frac{n!}{m_1! \cdot m_2! \cdot \ldots \cdot m_k!}. \qquad (7.9)$$

different ways of splitting a set containing $n \geq 0$ elements into $k \geq 0$ subsets containing $m_1, m_2, \ldots, m_k$ elements, respectively, with $m_1 + m_2 + \ldots + m_k = n$ and $m_i \geq 0$ for $i = 1, 2, \ldots, k$.

The 12 triples of sets listed in (7.8) illustrate this result. We had

$$\binom{4}{2 \ 1 \ 1} = \frac{4!}{2! \, 1! \, 1!} = 12.$$

7.8. Definition. The coefficient defined by Eq. (7.9) is called the *multinomial coefficient*. It generalizes the binomial coefficient of Theorem 7.1 and will be used in the next chapter in the definition of the 'multinomial distribution.'

The next theorem generalizes the Binomial Theorem (cf. 7.1).

The Multinomial Theorem

7.9. Theorem. *Let $a_1, a_2, \ldots, a_k$ be any k real numbers, and let n be a positive integer. Then*

$$(a_1 + a_2 + \ldots + a_k)^n = \sum_{(m_1, m_2, \ldots, m_k)} \binom{n}{m_1 \ m_2 \quad \ldots \quad m_k} a_1^{m_1} a_2^{m_2} \ldots a_k^{m_k},$$

where the summation runs over all k-tuples $(m_1, m_2, \ldots, m_k)$ of non-negative integers m_i, $0 \leq i \leq k$, satisfying $\sum_{i=1}^{k} m_i = n$.

PROOF. We proceed by induction on the value of k in the equation of the theorem. For $k = 1$, the result is trivial. By the Binomial Theorem 7.3, it also holds for $k = 2$. Suppose that, for any arbitary n it holds for some natural number $k = j > 2$, and consider the case $k = j + 1$. Using first the Binomial Theorem, and then the induction hypothesis, we obtain

$$(a_1 + a_2 + \ldots + a_{j+1})^n = (a_{j+1} + (a_1 + a_2 + \ldots + a_j))^n$$

$$= \sum_{m_{j+1}=0}^{n} \binom{n}{m_{j+1}} a_{j+1}^{m_{j+1}} (a_1 + \ldots + a_j)^{n-m_{j+1}}$$

$$= \sum_{m_{j+1}=0}^{n} \binom{n}{m_{j+1}} a_{j+1}^{m_{j+1}} \sum_{(m_1,\ldots,m_j)} \binom{n - m_{j+1}}{m_1 \ \ldots \ m_j} a_1^{m_1} \ldots a_j^{m_j}$$

$$= \sum_{(m_1,\ldots,m_{j+1})} \binom{n}{m_1 \ \ldots \ m_{j+1}} a_1^{m_1} \ldots a_{j+1}^{m_{j+1}},$$

because

$$\binom{n}{m_{j+1}} \binom{n - m_{j+1}}{m_1 \ \ldots \ m_j} = \binom{n}{m_1 \ \ldots \ m_{j+1}}. \qquad \square$$

7.10. Example. (RETURN TO THE CARD GAME EXAMPLE 6.1.) In this example, all the cards of a regular 52-card deck were randomly distributed among 4 players, each player receiving 13 cards. We asked: What is the probability that one of the players receives all the clubs? We should first notice an ambiguity in the question. Do we mean that a particular player, say George, ends up with all the clubs? Or do we mean that any player among the four receives all the clubs? Clearly, these are two closely related but different problems, and we shall consider both of them. We begin by numbering the four players: 1, 2, 3, and 4. The total number of points in the sample space is evidently, by Theorem 7.7,

$$\frac{52!}{13! \ 13! \ 13! \ 13!}.$$

To compute the probability that Player 1 (George) receives all the clubs, we 'freeze' these 13 club cards, and count the number of ways the other 39 cards can be evenly distributed among the other three players. Using Theorem 7.7 again, we obtain

$$\frac{39!}{13! \ 13! \ 13!}.$$

The probability that Player 1 receives all the clubs is thus the ratio

$$
\left(\frac{39!}{13!\,13!\,13!}\right) \Big/ \left(\frac{52!}{13!\,13!\,13!\,13!}\right) = \frac{39!\,13!}{52!}
$$

$$
\approx \frac{[(2\pi)^{\frac{1}{2}}39^{39+\frac{1}{2}}e^{-39}][(2\pi)^{\frac{1}{2}}13^{13+\frac{1}{2}}e^{-13}]}{(2\pi)^{\frac{1}{2}}52^{52+\frac{1}{2}}e^{-52}}
$$

$$
= \left(\frac{39\pi}{2}\right)^{\frac{1}{2}} \times \frac{3^{39}}{4^{52}}
$$

$$
\approx 1.56 \times 10^{-12}. \tag{7.10}
$$

In the second interpretation of this problem, some unspecified player among the four receives all the clubs. The probability of that event is approximately four times that given in the-right hand side of (7.10) because it is the disjoint union of the four events 'Player i gets all the clubs' with $1 \le i \le 4$.

7.11. Example. (RETURN TO THE BIRTHDAY PROBLEM 6.2.) What is the probability that among 20 people randomly gathered in a room, there are two of them with the same birthday? The question is ambiguous. Do we mean: 'exactly two people?' or 'at least two people?' We discuss the second interpretation and leave the first one to the reader (see Problem 2). Let B be the event that at least two people have the same birthday. We begin by computing (or estimating) the probability of the complementary event $\overline{B}$ that 'no two people have the same birthday.' As discussed in 6.2, let the sample space be the set of all 20-tuples $(x_1, \ldots, x_{20})$ with $1 \le x_i \le 365$ for $1 \le i \le 20$. The points in $\overline{B}$ are all the ordered samples of size 20, without replacement, in the set $\{1, 2, \ldots, 365\}$. The probability of B is thus

$$
\mathbb{P}(B) = 1 - \mathbb{P}(\overline{B}) = 1 - \frac{365_{20}}{365^{20}} = 1 - \frac{365!}{345!\,365^{20}}
$$

$$
\approx 1 - \frac{(2\pi)^{\frac{1}{2}}365^{365+\frac{1}{2}}e^{-365}}{(2\pi)^{\frac{1}{2}}345^{345+\frac{1}{2}}e^{-345}365^{20}}
$$

$$
= 1 - \left(\frac{73}{69}\right)^{345+\frac{1}{2}} e^{-20}
$$

$$
\approx .41\,.
$$

Note in passing that a much cruder paper-and-pencil method can be used to give a ballpark result. We write the probability of $\overline{B}$ more explicitly as

$$\mathbb{P}(\overline{B}) = \underbrace{\frac{365 \cdot 364 \cdot 363 \cdot \ldots \cdot 346}{365 \cdot 365 \cdot 365 \cdot \ldots \cdot 365}}_{20 \text{ factors}}$$

$$= (1 - \frac{1}{365})(1 - \frac{2}{365}) \ldots (1 - \frac{19}{365}).$$

Multiplying out the 19 factors in the last expression and neglecting in the resulting sum any term of the form $\pm(k/365^m)$ with $k > 1$ and $m > 1$, we get, after rearranging, the approximation

$$\mathbb{P}(\overline{B}) \approx 1 - \frac{1 + 2 + \ldots + 19}{365} = 1 - \frac{19 \cdot 20}{2 \cdot 365} = 1 - \frac{380}{730}. \qquad (7.11)$$

We obtain

$$\mathbb{P}(B) \approx \frac{380}{730} \approx .52 \,.$$

Problems

1. Give a conceptual proof, in the spirit of the proof of Theorem 7.4, for each of the following identities:

    (a) $\sum_{k=0}^{n} \binom{n}{k}^2 = \binom{2n}{n}$;

    (b) $\sum_{k=1}^{n} k\binom{n}{k} = n2^{n-1}$ (there also is a simple analytic proof of this identity via the Binomial Theorem; can you come up with it?);

    (c) $\sum_{i=0}^{k} \binom{m}{i}\binom{n}{k-i} = \binom{m+n}{k}$.

2. Solve the birthday problem for the first interpretation, that is, that there are exactly two people who share a birthday.

3. Consider a card game, using the standard 52-card deck, in which each of four players is dealt thirteen cards. Compute the probabilities that a specific player: (a) has no clubs; (b) has exactly ten clubs; (c) has at least three of the four aces.

4. (Continuation.) In the same card game, what is the probability that each player is dealt a jack?

5. Solve the following 'balls-into-cells' problems. (Note that the birthday problem is a balls-into-cells problem in which there are many more cells than balls.) Assume all the balls and all the cells are labeled.

(a) Find the probability that exactly one cell remains empty when n balls are placed at random into n cells.

(b) Find the probability that exactly two cells remain empty when n balls are placed at random into n cells. What is the probability that at least two cells are empty?

(c) Find the probability that at least two cells contain a ball.

6. Examine carefully the proof of the Multinomial Theorem and check the case $n = 3$ in detail.

7. (a) A researcher must choose seven subjects from each of six (disjoint) sets, with each set comprising fifteen people. In how many ways may she make her choices?

(b) A researcher must choose six subsets of seven people each from a set of 90 people. In how many ways may she make her choices?

8. A test taker may answer any m of n test questions $q_1, q_2, \ldots, q_n$.

(a) How many choices does he have?

(b) How many choices does he have if he must answer each of the first k questions?

(c) How many choices does he have if he must answer at least l of the first k questions?

9. Recall that there are 100 U.S. senators (two from each state).

(a) If two senators are chosen at random, what is the probability that they are from the same state?

(b) If the 100 senators are organized into disjoint sets of two, what is the probability that, in each set, the two senators are from the same state?

(c) In a committee of ten senators, what is the probability that no two are from the same state? (Assume that there are no restrictions on committee membership.)

10. A scientist wants to estimate the number of carps remaining in a lake after an ecological disaster. She catches 50 carps and tags them. Later on, she fishes again 50 carps and discovers that 10 of them are tagged. What is the probability of this event if the lake contains n carps? How can such data be used to estimate the number of carps remaining in the lake?

11. Ponder the relationship between Eq. (7.6) and the so-called 'Pascal's triangle' on the front cover of this book and also on p. v.

Chapter 8

Discrete Distributions

We recall that a probability distribution is a real valued function p defined on a finite or countable sample space Ω and satifying the two conditions: $p(x) \geq 0$ for all x in Ω, and $\sum_{x \in \Omega} p(x) = 1$ (cf. 4.5). Such a distribution is sometimes said to be *discrete*. Five cases of discrete distributions are considered in this chapter. In the first three cases, the sample space is finite.

The Bernoulli Distribution

8.1. Empirical Situation. An experimenter is selecting a single ball from an urn containing only black and white balls in proportions α and $1-\alpha$, respectively. When a black ball is selected, '1' is written on some experimental protocol; otherwise, '0' is recorded.

8.2. Definition. Let $\Omega = \{0, 1\}$ be a sample space, and let α be any real number, with $0 \leq \alpha \leq 1$. The probability distribution p defined on Ω by

$$p(x) = \begin{cases} \alpha & \text{if } x = 1 \\ 1 - \alpha & \text{if } x = 0 \end{cases} \tag{8.1}$$

is called a *Bernoulli distribution with parameter* α. Clearly, Eq. (8.1) defines a probability distribution on Ω. Moreover, any probability distribution on the sample space Ω is a Bernoulli distribution. (There is always some parameter $\alpha \geq 0$ satisfying (8.1).) Despite its simplicity, the Bernoulli distribution is fundamental because it enters in the definition or in the construction of many distributions

or processes. Notwithstanding the empirical situation described in
8.1, the parameter α need not be a rational number.

The Binomial Distribution

8.3. Empirical Situation. An experimenter is selecting n balls,
with replacement, from an urn containing only black and white balls.
The probability of getting a black ball if a single ball is selected is
equal to α. The number m of black balls obtained is recorded. In
some ideal conditions (see Remark 8.5(a)), the probability distribution defined below is an appropriate model for this situation.

8.4. Definition. Let $\Omega = \{0, 1, \dots, n\}$ be a sample space, with n
a positive integer, and let α be any real number, with $0 \leq \alpha \leq 1$.
The probability distribution p defined on Ω by

$$p(m) = \binom{n}{m} \alpha^m (1 - \alpha)^{n-m} \tag{8.2}$$

is called a *binomial distribution with parameters α and n*. Note
that the three factors in the right hand side of (8.2) are nonnegative, yielding $p(m) \geq 0$ for $m = 0, 1, \dots, n$. Also, by the Binomial
Theorem 7.3, we have

$$\sum_{m=0}^{n} p(m) = \sum_{m=0}^{n} \binom{n}{m} \alpha^m (1 - \alpha)^{n-m} = [\alpha + (1 - \alpha)]^n = 1.$$

We conclude that the function p defined by (8.2) is indeed a probability distribution.

8.5. Remarks. (a) It can be proven that, under certain conditions
concerning the independence of the successive drawings of the balls,
the distribution defined by (8.2) is indeed that required for the empirical situation described in 8.3.

(b) By definition of the binomial distribution p with parameters
$\alpha = \frac{1}{2}$ and n, we have (using 7.4, Eq. (7.5)),

$$p(m) = \binom{n}{m} \left(\frac{1}{2}\right)^m \left(1 - \frac{1}{2}\right)^{n-m}$$

$$= \binom{n}{n-m} \left(\frac{1}{2}\right)^{n-m} \left(1 - \frac{1}{2}\right)^m = p(n - m).$$

This equality holds for every point m of the sample space. To express this fact, we sometimes say that the binomial distribution with parameters $\frac{1}{2}$ and n is *symmetric around* $\frac{n}{2}$. If n is even, then $\frac{n}{2}$ is a point of the sample space, and $p(\frac{n}{2})$ is a maximum of the function $p(n)$. If n is odd, then the function $p(n)$ has two equal maxima located at the two contiguous values $\frac{n-1}{2}$ and $\frac{n+1}{2}$. (You are asked to prove this in Problem 2.)

The Multinomial Distribution

8.6. Empirical Situation. An experimenter is sampling balls with replacement, from an urn containing k different types of balls, numbered 1, ... , k. If a single ball is selected from the urn, the probability that it is a ball of type i ($1 \le i \le k$) is equal to α_i; thus, $\sum_{i=1}^{k} \alpha_i = 1$. Suppose that n balls are sampled, and that the experimenter records the numbers m_i ($0 \le i \le k$) of balls of each type, with $\sum_{i=1}^{k} m_i = n$. Under certain conditions regarding the independence of the successive drawings of the balls,[1] the distribution defined below is a suitable model for this situation.

8.7. Definition. Let n and k be two integers, with $n \ge k > 0$. Let the sample space Ω be the set of all k-tuples $(m_1, \dots, m_k)$ of nonnegative integers satisfying $\sum_{i=1}^{k} m_i = n$. Let $\alpha_1, \dots, \alpha_k$ be nonnegative real numbers satisfying $\sum_{i=1}^{k} \alpha_i = 1$. The probability distribution p defined on Ω by the equation

$$p(m_1, m_2, \dots, m_k) = \begin{pmatrix} & & n & \\ m_1 & m_2 & \dots & m_k \end{pmatrix} \alpha_1^{m_1} \alpha_2^{m_2} \dots \alpha_k^{m_k} \quad (8.3)$$

is called the *multinomial distribution with parameters* $\alpha_1, \dots, \alpha_k$ and n. In the situation described in 8.6, a value $p(m_1, \dots, m_k)$ of this distribution describes (in terms of the parameters $\alpha_1, \dots, \alpha_k$ and n) the probability of sampling exactly m_1 balls of type 1, m_2 balls of type 2, ... , m_k balls of type k, with $\sum_{i=1}^{k} m_i = n$. To check that the function p defined by (8.3) is indeed a probability distribution, observe that all the factors in (8.3) are nonnegative, and apply the Multinomial Theorem 7.9. In the case $k = 2$, the multinomial distribution reduces to the binomial distribution.

---

[1] Essentially, the drawing of a ball and replacing it in the urn does not affect the likelihood of drawing any particular ball afterwards.

The two probability distributions remaining to be discussed in the rest of this chapter are defined on a countable sample space, specifically, either the set $\mathbb{N}$ of natural numbers or the set $\{0, 1, \dots, n, \dots\}$.

The Geometric Distribution

8.8. Empirical Situation. An experimenter is drawing balls from an urn containing black and white balls, replacing the balls in the urn after each selection. We suppose that the probability of drawing a black ball is a constant α $(0 < \alpha < 1)$. The drawing is continued until the first black ball is drawn. The experimenter only records the number of the particular trial where this occurs. Thus, if the first black ball is drawn on trial 5, the experimenter records the number 5 as the outcome of the experiment. (A similar example was discussed in 2.2, in which a coin was tossed until the first head occurred.)

8.9. Definition. Let $\Omega = \{1, 2, \dots, n, \dots\}$ be the sample space, and let α be a real number with $0 < \alpha < 1$. The function p defined on Ω by the equation

$$p(n) = \alpha(1 - \alpha)^{n-1} \qquad\qquad (8.4)$$

is called a *geometric distribution with parameter* α. The function p is clearly nonnegative and we have

$$\sum_{n=1}^{\infty} p(n) = \alpha \left(1 + (1 - \alpha) + (1 - \alpha)^2 + \dots + (1 - \alpha)^n + \dots\right) = 1.$$

As specified by (8.4), the function p satisfies thus the two defining conditions of a probability distribution.

The Poisson Distribution

8.10. Empirical Situation. An experimenter is watching a display for the appearance of signals of a special kind. We suppose that these signals have the following characteristics: they are punctual (that is, they have no duration); there is no refractoriness in the system delivering the signals (that is, the fact that a signal has just been detected does not carry any information regarding the occurence of the next one); and the occurrence of the signals is random but homogeneous, in the sense that the signal is just as likely to occur within one interval of time of a given length as within some other interval of the same length. The experimenter watches the display for a fixed duration, which we arbitrarily set at one unit of time, and counts the number of signals occurring during that time. Intuitively, this situation seems to be a continuous analogue of the empirical situation described in 8.3, in which an experimenter was counting the number of black balls drawn in a sequence of n trials. The distribution there was the binomial distribution, and it is indeed possible to derive the Poisson distribution, which is the subject of this section, as a limiting case of the binomial distribution.

8.11. Derivation of the Poisson Distribution. We begin by reshaping the situation described in 8.10 into one to which the binomial distribution might be applicable. We divide the observation time (whose length was supposed to be equal to one) into n subintervals of equal length $[0, t_1],]t_1, t_2], \dots,]t_{n-1}, t_n]$. We call E_1 the event that 'at least one signal was detected during the first time interval $[0, t_1]$.' Similarly, we denote by $E_2, \dots, E_n$ the events that 'at least one signal was detected during the subintervals $]t_1, t_2], \dots,]t_{n-1}, t_n]$,' respectively. Because all the n subintervals have the same duration, and in view of the informal assumption of homogeneity of the occurrence of the signals, it is natural to assume that the probabilities of all the events E_i is the same. It is also natural to suppose that the probability of any event E_i is proportional to the duration of its associated interval, which by convention is equal to $\frac{1}{n}$. Let us denote by $\frac{\lambda}{n}$ the probability that E_i is realized, with $1 \leq i \leq n$. The number k of events E_i that can be realized through one outcome (one unit time observation) ranges between 0 and n. Invoking again the assumptions of 8.10, it is reasonable to postulate that k is distributed binomially with parameters $\frac{\lambda}{n}$ and n. We obtain, for the probability $p_n(k)$ of observing k events E_i, the equation

$$
p_n(k) = \binom{n}{k} \left(\frac{\lambda}{n}\right)^k \left(1 - \frac{\lambda}{n}\right)^{n-k}
$$

$$
= \frac{n!}{(n-k)! \, n^k} \left(1 - \frac{\lambda}{n}\right)^{-k} \left(1 - \frac{\lambda}{n}\right)^n \frac{\lambda^k}{k!}
$$

(after factoring out the terms not depending on n). We rewrite the last equation as

$$p_n(k) = \left(1 - \frac{1}{n}\right)\left(1 - \frac{2}{n}\right)\cdots\left(1 - \frac{k-1}{n}\right)\left(1 - \frac{\lambda}{n}\right)^{-k}\left(1 - \frac{\lambda}{n}\right)^n \frac{\lambda^k}{k!}.$$

Letting $n \to \infty$, all but the last two factors tend to 1, and we know from calculus that $[1 - (\lambda/n)]^n$ tends to $e^{-\lambda}$. We obtain finally

$$\lim_{n \to \infty} p_n(k) = e^{-\lambda}\frac{\lambda^k}{k!}.$$

8.12. Definition. Let $\Omega = \{0, 1, \dots, n, \dots\}$ be the sample space, and let λ be a positive real number. The probability distribution p defined on Ω by the equation

$$p(k) = e^{-\lambda}\frac{\lambda^k}{k!} \tag{8.5}$$

is called a *Poisson distribution with parameter* λ. This equation defines a probability distribution because the three factors in the right-hand side are nonnegative, and we have

$$\sum_{k=0}^{\infty} p(k) = e^{-\lambda} \times \left(1 + \frac{\lambda}{1!} + \frac{\lambda^2}{2!} + \frac{\lambda^3}{3!} + \dots\right) = e^{-\lambda}e^{\lambda} = 1.$$

As suggested by the discussion preceding Definition 8.12, the Poisson distribution can be taken to approximate the binomial distribution. This approximation is very good in some situations (for instance, the approximation of a binomial distribution with parameters n and $\frac{1}{n}$ by a Poisson distribution with parameter $\lambda = 1$.)

The Poisson distribution is one aspect of a stochastic process called the 'Poisson process' which describes mathematically the occurrence in time of punctual events according to the principles outlined in 8.11. This is the topic of Chapter 24. For an introduction to the theory of stochastic processes, see for example Parzen (1994b) or Barucha-Reid (1974). Chapters 21 to 24 of this monograph are devoted to some examples of stochastic processes, namely: Markov chains (Chapters 21 and 22), random walks (Chapter 23), and Poisson processes (Chapter 24).

Problems

1. Tabulate the binomial distribution in two cases: (a) $n = 5$ and $\alpha = \frac{1}{3}$;
   (b) $n = 5$ and $\alpha = \frac{1}{2}$.

2. For a binomial distribution with parameters α and n, prove that if $\alpha = \frac{1}{2}$
   and n is even, then $p(\frac{n}{2})$ is a maximum of the function $p(n)$. Prove also
   that if n is odd, then the function $p(n)$ has two equal maxima located at
   the two contiguous values $\frac{n-1}{2}$ and $\frac{n+1}{2}$.

3. Compute the probability that, in a sample of 10 people in the population
   at large, 2 have their birthdays in May or June, 3 have their birthdays
   in December or January, and the 5 remaining ones have their birthdays
   during the rest of the year. (Just give the formula. Also, you may assume
   for simplicity that all the months have the same number of days.)

4. Two fair dice are thrown repeatedly until for the first time their sum
   exceeds 4. What is the distribution of the trial number of that event?

5. Check the accuracy of the approximation of the binomial distribution with
   parameters 50 and $\frac{1}{50}$ by the Poisson distribution with parameter 1. Compute both distributions for the values 0, 1, 3, and 5.

6. For each of the phenomena described below, propose a probability distribution for the numerical variable(s) involved and give the corresponding
   formula.

   (a) A hospital specializes in the treatment of five types of cancer, la-
       beled $C_1, \dots, C_5$, which are know to occur in the population with
       respective probabilities $\alpha_1, \dots, \alpha_5$. On a particular year, 200 pa-
       tients are admitted in the hospital. Compute the probability that,
       for $1 \leq i \leq 5$, there are n_i patients suffering from disease C_i. (You
       may assume that no patient has more than one disease.)

   (b) In the same hospital, the long-term probability of recovery from dis-
       ease C_1 is .85. Suppose that 50 patients with that disease are ad-
       mitted in the hospital on a particular year. Compute the probability
       that k of them recover.

   (c) Suppose that a couple has k children. Compute the probability that
       there is at least one girl among them. In this and in Problem 6(d),
       you may assume that the probability of having a girl is α.

   (d) In the same situation (that is, a couple with k children), compute
       the probability that the number of boys equals the number of girls.

   (e) Compute the probability of observing exactly k alternations of colors
       in a drawing of 6 balls, with replacement, from an urn containing
       an equal number of red and blue balls. (Examples: RBBBBB is
       one alternation, whereas RBBRBB and RRBRBB each have three
       alternations.)

7. A couple decides to have at most 5 children, but to stop as soon as two girls are born. What is the probability that the family end up with 4 children? You may assume that the probability of having a girl is .48.

8. In a particular ESP (extrasensory perception) experiment, an experimenter looks at one of five cards on each trial, and the subject is to guess at which card the experimenter is looking. Assuming that the subject does not have ESP, what is the distribution of the outcome (successful guess, unsuccessful guess) of a trial?

9. (Continuation.) Suppose that twenty trials of the above experiment are performed and that the subject does not have ESP. (a) What is the distribution of the number of successful responses by the subject in those twenty trials? (b) What is the probability that the subject is successful on at least four trials?

10. (Continuation.) Repeated trials of the above experiment are performed until the subject guesses correctly. (a) What is the distribution of the trial number of this event? (b) What is the probability that this event occurs on the fifth trial? (c) What is the probability that this event occurs before the fifth trial?

11. In the empirical situation of Example 8.8, suppose that the experimenter proceeds until she draws r black balls. (a) Write an expression for the probability $p_1(n)$ that she draws exactly n balls total. (b) Show that p_1 is a probability distribution. (It is called the *negative binomial* distribution.)

12. (Continuation.) Suppose now that the urn contains N balls, of which M are black. (a) Write an expression for the probability $p_2(i)$ that, in n draws WITHOUT REPLACEMENT, the experimenter obtains exactly i black balls. (b) Show that p_2 is a probability distribution. (It is called the *hypergeometric* distribution.) Hint: Use an identity appearing in the Chapter 7 Problems.

*13. This is an elaboration of Problem 6(b). Suppose that, in the same hospital, the long-term probability of recovery from disease C_i is ξ_i, for $1 \leq i \leq 5$. What is the probability that n patients will recover among the 200 patients admitted on a particular year?

*14. Roger and Emile decide to gamble on a coin tossing game. Roger, who knows that the coin is biased and that the probability of a tail is only .4, chooses 'head.' Both start the game with $5. On each toss, the winner receives one dollar from the loser. They decide to toss the coin 10 times (or fewer, if one of the two players runs out of money). What is the probability of the ruin of Emile? (Just give the formula.)

Chapter 9

Conditional Probabilities

Intuitively, the conditional probability of some event A 'given' an event B is the probability that A occurs when we know for sure that B must also occur (or has occurred). The usual notation is $\mathbb{P}(A \mid B)$. As a preparation for a formal definition, we illustrate this concept by an example.

Examples and Definition

9.1. Example. What is the probability that two successive tosses of a die give a total of six given that both numbers are even? Let us examine the situation in some detail. The sample space for this experiment is given by Table 9.1, in which each cell represents an outcome, namely, a possible result for the two tosses. The event 'total is six' is marked by '★,' while 'both numbers are even' is represented by '○.'

Table 9.1: Sample space of the 'two tosses of a die' experiment, with the two events: (i) both numbers are even (○); (ii) the total is 6 (★).

	1	2	3	4	5	6
1	—	—	—	—	★	—
2	—	○	—	⊛	—	○
3	—	—	★	—	—	—
4	—	⊛	—	○	—	○
5	★	—	—	—	—	—
6	—	○	—	○	—	○

Assuming that the die is fair, all 36 outcomes have the same probability. Table 9.1 suggests that the probability of obtaining a total equal to 6 given that both numbers are even should be equal to 2/9: indeed, there are two cases 'total is 6' among the nine cases 'both numbers are even.' This result was obtained by an intuitive analysis of the situation. We shall now examine the computation in more detail.

Let us denote by '$\mathbf{S} = 6$' the event 'the total is 6.' (Thus, '$\mathbf{S} = 6$' is a compact notation for the event $\{(1,5),(2,4),...,(5,1)\}$.) We also denote by '$\mathrm{E}$' the event 'both numbers are even.' Stating that the coin is fair indicates that the counting measure should be used. We have thus

$$\mathbb{P}(\mathrm{E}) = \frac{|\mathrm{E}|}{|\Omega|} = \frac{9}{36},$$

$$\mathbb{P}\left((\mathbf{S} = 6) \cap \mathrm{E}\right) = \frac{|(\mathbf{S} = 6) \cap \mathrm{E}|}{|\Omega|} = \frac{2}{36},$$

and the value of 2/9 for the conditional probability of the event $(\mathbf{S} = 6)$ given the event E results from the computation

$$\mathbb{P}(\mathbf{S} = 6 \,|\, \mathrm{E}) = \frac{\mathbb{P}[(\mathbf{S} = 6) \cap \mathrm{E}]}{\mathbb{P}(\mathrm{E})} = (2/36)/(9/36) = 2/9.$$

Loosely speaking, one interpretation of this analysis is that, to compute $\mathbb{P}(A \,|\, B)$, one has to assign a probability of one to the event B, and to 'normalize' the probability of the event A accordingly. Let us consider another example, using the geometrical representation that we are familiar with.

9.2. Example. Consider the case of three events, A, B and C in a sample space Ω. The sample space is represented by a rectangular region (see Fig. 9.1). The events A, B and C are represented by the regions surrounded by the three 'squarish' ovals. The probabilities assigned to all the regions are marked or can easily be computed. For instance, the probability of $\overline{(A \cup B \cup C)}$ is .10, and the probability of the event $\overline{(B \cup C)}$ is .35 ($= .25 + .10$). Let us compute the conditional probability of the event A given the event $B \cup C$, that is $\mathbb{P}(A \,|\, B \cup C)$. This suggests neglecting all the regions outside $B \cup C$, as indicated by Fig. 9.2, in which all those probabilities have been deleted. The remaining probabilities are in parentheses, to indicate that these numbers have to be normalized. We have

$$\mathbb{P}(B \cup C) = .05 + .05 + .05 + .30 + .10 + .10 = .65.$$

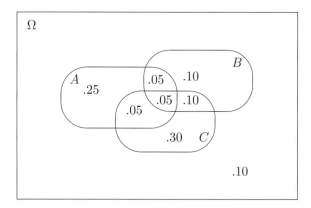

Figure 9.1: The oval regions corresponding to the events A, B, **and** C, **with the probabilities associated with all the subregions of** $A \cup B \cup C$. **Note that** $\mathbb{P}(\overline{A \cup B \cup C}) = .10$.

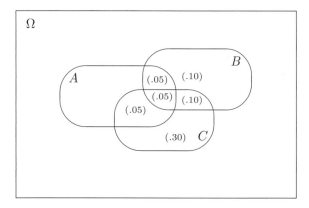

Figure 9.2: The union of the two regions with no probabilities assigned forms the complement of the conditioning event $B \cup C$. **The probabilities in the parentheses have to be normalized.**

For computing the conditional probability $\mathbb{P}(A \mid B \cup C)$, we proceed as if $B \cup C$ were assigned a probability equal to 1, and obtain

$$\mathbb{P}(A \mid B \cup C) = \frac{.05 + .05 + .05}{.65} = \frac{\mathbb{P}[A \cap (B \cup C)]}{\mathbb{P}(B \cup C)}.$$

Note in passing that, using this rule, we get

$$\mathbb{P}(B \cup C \mid B \cup C) = \frac{\mathbb{P}[(B \cup C) \cap (B \cup C)]}{\mathbb{P}(B \cup C)} = 1.$$

These examples lead to the following general definition.

9.3. Definition. Let $(\Omega, \mathcal{F}, \mathbb{P})$ be a finitely additive probability space. For all events A, B such that $\mathbb{P}(B) \neq 0$, we define

$$\mathbb{P}(A \mid B) = \frac{\mathbb{P}(A \cap B)}{\mathbb{P}(B)}. \tag{9.1}$$

The left-hand side of (9.1) is referred to as the *conditional probability* of A *given* B. The reason for the condition $\mathbb{P}(B) \neq 0$ is obviously that the right side of (9.1) is not defined if $\mathbb{P}(B) = 0$. The conditional probability of an event A given an event B is thus only defined if $\mathbb{P}(B) \neq 0$. As suggested by the example below, this is a severe limitation.

9.4. Example. Suppose that a medical researcher is interested in the relation between heart attacks and cholesterol level. For concreteness, we denote by H the event of observing at least one heart attack in a particular year, and by $\mathbf{L} = x$—where x is a positive real number—the event that the cholesterol level, measured at the beginning of the year, is exactly equal to x. Thus, two typical sample points are (H,x) and $(\overline{H}, x)$, where $\overline{H}$ denotes the event: 'no heart attack observed during the year.' One natural possibility is to investigate the conditional probability that a heart attack is observed in a particular year given a cholesterol level equal to x, that is,

$$\mathbb{P}(H \mid \mathbf{L} = x). \tag{9.2}$$

Under most reasonable theories for such a situation, the probability that the cholesterol level of an individual is equal to x (for any particular number x), is equal to zero. This means that, under the definition of conditional probability given above, the quantity in (9.2) is not defined. In some more advanced presentations of probability theory, a definition of conditional probability is given which generalizes (9.1) and allows quantities such as (9.2) to be defined. (See Parzen, 1994a, p. 334-336, for additional comments.)

Some Consequences

Let $(\Omega, \mathcal{F}, \mathbb{P})$ be a finitely additive probability space. Intuitively, when we condition by some event $C \in \mathcal{F}$, with $C \neq \Omega$, it is (almost) as if we were restricting consideration to a smaller sample space $C \subseteq \Omega$. For example, suppose that $A \cap C$ and $B \cap C$ are two incompatible events, and that $\mathbb{P}(C) \neq 0$. It is not difficult to show that

$$\mathbb{P}(A \cup B \mid C) = \mathbb{P}(A \mid C) + \mathbb{P}(B \mid C),$$

which is essentially the defining (finite) additivity property of probabilities[1] holding for the smaller space C. This idea will be made precise in a moment. A preliminary result is needed.

9.5. Theorem. *Suppose that $(\Omega, \mathcal{F}, \mathbb{P})$ is a finitely additive probability space. For all events A, B, and C, such that $\mathbb{P}(C) \neq 0$, we have*

(i) $\mathbb{P}(\Omega \mid C) = 1$;

(ii) $\mathbb{P}(A \mid C) \geq 0$;

(iii) *if $A \cap B = \emptyset$, then* $\mathbb{P}(A \cup B \mid C) = \mathbb{P}(A \mid C) + \mathbb{P}(B \mid C)$;

(vi) *if $A \cap C = \emptyset$, then* $\mathbb{P}(A \mid C) = 0$;

(v) *if* $\mathbb{P}(A) \neq 0$, *then* $\mathbb{P}(A \mid C) = \frac{\mathbb{P}(A)\mathbb{P}(C \mid A)}{\mathbb{P}(C)}$.

Each of (i) to (v) follows immediately from the definition of conditional probability. The proofs are left to the student (Problem 6).

9.6. Theorem. *Suppose that $(\Omega, \mathcal{F}, \mathbb{P})$ is a finitely additive probability space. Let C be some event such that $\mathbb{P}(C) \neq 0$. For any $A \in \mathcal{F}$, define $\mathbb{P}_C(A) = \mathbb{P}(A \mid C)$. Then $(\Omega, \mathcal{F}, \mathbb{P}_C)$ is a finitely additive probability space.*

PROOF. Since we already know that $\mathcal{F}$ is a field of events, we simply have to show that $\mathbb{P}_C$ satisfies Axioms [K1], [K2] and [K3] of a finitely additive probability space. These conditions are, respectively, parts (i), (ii), and (iii) of the preceding theorem.   □

We also indicate the following useful result, the proof of which is also left to the student (Problem 7).

---

[1] See Kolmogorov's Axiom [K3] of Def. 4.1.

9.7. Theorem. *For any finite collection of events $A_1, \dots , A_n$, we have*

$$\mathbb{P}(A_1 \cap A_2 \cap \dots \cap A_n) =$$
$$\mathbb{P}(A_1)\, \mathbb{P}(A_2 \,|\, A_1)\, \mathbb{P}(A_3 \,|\, A_1 \cap A_2) \dots \mathbb{P}(A_n \,|\, A_1 \cap \dots \cap A_{n-1}).$$

The Theorem of Total Probabilities

The next result is extremely important, both from a theoretical and practical viewpoint. The idea is that the probability of an event A can in many cases be decomposed additively in terms of the probabilities of other events that 'cover' Ω. This type of analysis may be useful whenever the direct computation of event probabilities appears difficult.

A graphic representation of a general situation may be helpful. In Fig. 9.3, an event A is pictured by a squarish oval region. The sample space Ω is represented by the rectangular region surrounded by the frame of the figure, and there are also events $H_1, H_2, \dots , H_5$, covering Ω, pairwise incompatible and represented by the five polygonal regions that are covering the rectangle.

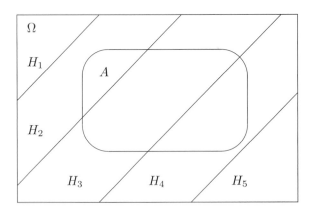

Figure 9.3: The event A and the covering of Ω by the disjoint events $H_1, \dots , H_5$.

The idea is to decompose the probability of A through the sum

$$\mathbb{P}(A) = \mathbb{P}(A \cap H_1) + \mathbb{P}(A \cap H_2) + \mathbb{P}(A \cap H_3) + \mathbb{P}(A \cap H_4) + \mathbb{P}(A \cap H_5)$$

which, if $\mathbb{P}(H_i) > 0$ for $1 \leq i \leq 5$, can also be written as

$$\mathbb{P}(A) = \sum_{i=1}^{5} \mathbb{P}(A \mid H_i)\mathbb{P}(H_i).$$

In the case of the figure, the first and the last terms in these two sums vanish because the corresponding intersections are empty, but the decomposition is nevertheless valid. In the next theorem, we consider the general situation.

We recall that, when the events $E_1, E_2, \ldots, E_n$ are pairwise incompatible, then the notation $\sum_{i=1}^{n} E_i$ has the same meaning as the notation $\cup_{i=1}^{n} E_i$. The finite version of the *Theorem of Total Probabilities* is formulated below.

9.8. Theorem. *Let $(H_i)_{1 \leq i \leq n}$ be a family of pairwise incompatible events in a finitely additive probability space $(\Omega, \mathcal{F}, \mathbb{P})$, such that $\sum_{i=1}^{n} H_i = \Omega$. Then, for all $A \in \mathcal{F}$,*

$$\mathbb{P}(A) = \sum_{i=1}^{n} \mathbb{P}(A \cap H_i). \tag{9.3}$$

Moreover, if $\mathbb{P}(H_i) \neq 0$, for all i, $1 \leq i \leq n$, then,

$$\mathbb{P}(A) = \sum_{i=1}^{n} \mathbb{P}(A \mid H_i)\mathbb{P}(H_i). \tag{9.4}$$

PROOF. If $\mathbb{P}(H_i) \neq 0$, we have $\mathbb{P}(A \cap H_i) = \mathbb{P}(A \mid H_i)\mathbb{P}(H_i)$, by definition of the conditional probability. Thus, (9.3) implies (9.4). Consequently, the theorem holds if we establish (9.3). Successively:

$$\mathbb{P}(A) = \mathbb{P}(A \cap \Omega)$$

$$= \mathbb{P}\left(A \cap \left(\sum_{i=1}^{n} H_i\right) \right) \qquad \text{(by hypothesis)}$$

$$= \mathbb{P}\left(\sum_{i=1}^{n} (A \cap Hi) \right) \qquad \text{(by set theory)}$$

$$= \sum_{i=1}^{n} \mathbb{P}(A \cap H_i) \qquad \text{(by finite additivity).}$$

$\square$

9.9. Remark. Versions of this theorem also hold in the cases of countable and uncountable collections of events H_i. The probabilities of the joint events $A \cap E_i$ in (9.3) are then replaced by a joint density function (cf. Chapters 19 and 20).

Problems

1. Archie has three coins in his pocket: a standard coin, a coin with heads on both sides, and a coin with tails on both sides. He pulls one coin out of his pocket, looks at one side of the coin, and notices that it is a tail. He reasons that the probability of seeing a head on the other side of this coin is $\frac{1}{2}$. Do you agree with his reasoning?

2. Let a, b, and c be outcomes in some finite sample space Ω having 2^Ω as a field of events, with some probability measure $\mathbb{P}$. You are told that $\mathbb{P}(\{a,b\} \,|\, \{b,c\}) = \alpha$ and that $\mathbb{P}(\{c\}) = \beta$.

   (a) Compute $\mathbb{P}(\{b\})$ in terms of α and β.

   (b) Give some possible values for α and β.

   (c) Find constraints on α and β, that is, find a general expression[2] constraining the possible values of α and β.

   (d) Find an expression constraining the possible values of $\mathbb{P}(\{a\})$.

3. Five prisoners A, B, C, D, and E are informed that three of them are going to be released. Prisoner B is told that either C or D is going to be released. What is the probability that prisoner B is going to be released, assuming that the guard is telling the truth?

4. There are three hats on the table: two green and one red. There are also two ties in the closet: one green and one red. Each of two brothers, Ralph and Rudolph, is randomly picking one hat and one tie.

   (a) What is the probability that each of the two brothers has matching colors?

   (b) What is the conditional probability that Rudolph has a green hat, given that Ralph has matching colors?

5. Archie must travel to a foreign country where an epidemic of a deadly strain of Asian flu has struck. About 17% of the people in large urban centers have died already. The chances of dying, for a nonvaccinated person in this country, have been evaluated at 1/5. Due to the fast development of the disease, only 20% of the people have been vaccinated. Archie wants to know (as accurately as possible) what his chances of survival are if he undergoes vaccination before departure. Can you help him?

---

[2]In the form of inequalities. Obviously, a justification must be provided.

6. Prove the results (i)-(v) in Theorem 9.5.

7. Provide a proof of Theorem 9.7.

8. (Chung (1974).) Consider, in some ethnic group, the set of all families having two children. Let us assume that, for such families, the probability of having a boy is .5. Take one family at random and suppose that one of their children is a boy. What is the probability that the other child is also a boy?

9. Archie is working for Life Magazine and has been given the assignment to go to the Galápagos Islands and photograph a very rare giant white turtle. Two possible sites have been mentioned: the turtle can be found either on Isabella (the largest of the Galápagos Islands), or Santa Cruz (the next largest of these islands[3]). Many explorers have attempted to photograph the turtle, but 92% have failed to locate it. Most people (80%, actually) have tried to find the turtle on Santa Cruz, and 5% of them have been successful. Archie has a hunch that he will find the turtle on Isabella. Would he be right to try Isabella? What is the probability that he would be successful?

10. Mr. Fitzwilliam Darcy just bought a house and moved to a wealthy neighborhood. He then learned that 10% of the houses had been burglarized during the preceding year. Wishing to avoid a similar fate, Mr. Darcy inquires about available burglar alarm systems. Two of them, A and B, are used in approximately equal proportions by his neighbors, and the claim is that they also offer equal protection. On the other hand, one fifth of the houses have no alarm system, and 40% of those have been burglarized. Mr. Darcy, who possesses expensive original art, wants to know what his chances are to be burglarized this coming year, were he to acquire one of the two systems. Can you help him?

11. Write a formula for the probability that, in a bridge game, George receives exactly ten spades, given that his partner Rudolpho receives all the hearts.

12. An astronomer has detected punctual signals from an unknown source in the sky. The signals are of two kinds, which she denotes 'A' and 'B.' She assumes that the occurrence of the signals is governed by a random process, namely, that the number of signals of any kind received in the course of one hour has a Poisson distribution with parameter λ. When a signal occurs, it is an 'A' signal with probability θ and a 'B' signal with probability $1 - \theta$. Write a formula for the distribution of the number of 'A' signals received in one hour.

---

[3]There are 16 islands in the Galápagos Archipelago.

Chapter 10

Independence and Bayes Theorem

Independence of Events

Loosely speaking, an event A is 'independent' of some event B when knowing that B is realized does not affect the probability of A. Formally, this idea is expressed by the equation:

$$\mathbb{P}(A \mid B) = \mathbb{P}(A).\qquad(10.1)$$

As a definition, Eq. (10.1) is not satisfactory because it only applies when $\mathbb{P}(B) > 0$. The standard definition below is more general and implies (10.1) whenever $\mathbb{P}(B) > 0$ (see Theorem 10.4 (ii)).

10.1. Definition. In a finitely additive probability space $(\Omega, \mathcal{F}, \mathbb{P})$, two events A and B are called *independent* if

$$\mathbb{P}(A \cap B) = \mathbb{P}(A)\mathbb{P}(B).\qquad(10.2)$$

It is not required that $\mathbb{P}(A), \mathbb{P}(B) > 0$. However, if $\mathbb{P}(A) = 0$, then $\mathbb{P}(A \cap B) = 0$ (since $A \cap B \subseteq A$), and $\mathbb{P}(A \cap B) = \mathbb{P}(A)\mathbb{P}(B) = 0$. An event of probability zero is thus independent of any other event.

10.2. Example. We go back to the dice tossing experiment discussed in 9.1 and consider the two events

A: 'odd face with the first die'

B: 'odd face with the second die.'

We have thus:

$$A = \{(1,1), \ldots, (1,6), (3,1), \ldots, (3,6), (5,1), \ldots, (5,6)\},$$
$$B = \{(1,1), \ldots, (6,1), (1,3), \ldots, (6,3), (1,5), \ldots, (6,5)\},$$
$$A \cap B = \{(1,1), (1,3), (1,5), \ldots, (5,3), (5,5)\}.$$

This implies

$$\mathbb{P}(A) = \mathbb{P}(B) = \frac{18}{36} \quad \text{and} \quad \mathbb{P}(A \cap B) = \frac{9}{36}.$$

Hence, $\mathbb{P}(A \cap B) = \mathbb{P}(A)\mathbb{P}(B) = \frac{1}{4}$. (If in doubt, you should go back to Table 9.1 and count.)

10.3. Remark. It is of some interest to notice that three events A, B, and C can be pairwise independent, while $A \cap B$ and C are not. For instance let C be the event 'odd sum' in the above example. It is easy to compute that $\mathbb{P}(C) = \frac{1}{2}$. But $A \cap B$ and C are incompatible; hence

$$\mathbb{P}(A \cap B \cap C) = \mathbb{P}(\emptyset) = 0 \neq \mathbb{P}(A \cap B)\mathbb{P}(C) = \frac{1}{4} \times \frac{1}{2}.$$

(This example is from Feller, 1957, p. 116.)

We end this section with a result giving some of the consequences of Definition 10.1.

10.4. Theorem. *Suppose that A and B are two independent events in a finitely additive probability space $(\Omega, \mathcal{F}, \mathbb{P})$. Then*

(i) *$\overline{A}$ and B are independent;*

(ii) *if $\mathbb{P}(B) \neq 0$, then $\mathbb{P}(A \mid B) = \mathbb{P}(A)$.*

Part (ii) of this theorem justifies the adjective 'independent' qualifying two events A and B satisfying Eq. (10.2)

PROOF. (i) Successively,

$$\begin{aligned}
\mathbb{P}(\overline{A} \cap B) &= \mathbb{P}(B) - \mathbb{P}(A \cap B) && \text{(Theorem of Total Probabilities)} \\
&= \mathbb{P}(B) - \mathbb{P}(A)\mathbb{P}(B) && \text{(A and B are independent)} \\
&= \mathbb{P}(B)[1 - \mathbb{P}(A)] \\
&= \mathbb{P}(B)\mathbb{P}(\overline{A}).
\end{aligned}$$

(ii) By Definitions 9.3 and 10.1, with A and B independent,

$$\mathbb{P}(A \mid B) = \frac{\mathbb{P}(A \cap B)}{\mathbb{P}(B)} = \frac{\mathbb{P}(A)\mathbb{P}(B)}{\mathbb{P}(B)} = \mathbb{P}(A).$$

$\square$

Bayes Theorem

The celebrated result known as Bayes Theorem is an immediate consequence of the Theorem of Total Probabilities (Theorem 9.8). The fame of this theorem should not be attributed to its mathematical profundity. Rather, it is due to its importance in applications and also to the philosophical issues that have been raised in its context. We shall first state and prove Bayes Theorem. Next, we shall consider some of its philosophical underpinnings and illustrate its application by examples. In Chapter 11, we shall discuss the Principle of Maximum Likelihood, which is a natural by-product of Bayes Theorem.

10.5. Theorem. (BAYES THEOREM.) *Let* $(\mathbf{H}_i)_{1 \leq i \leq n}$ *be a family of pairwise incompatible events in a finitely additive probability space* $(\Omega, \mathcal{F}, \mathbb{P})$, *satisfying* $\cup_{i=1}^{n} \mathbf{H}_i = \Omega$ *and* $\mathbb{P}(\mathbf{H}_i) \neq 0$, *for all* i, $1 \leq i \leq n$. *If* $D \in \mathcal{F}$ *is any event such that* $\mathbb{P}(D) \neq 0$, *then, for* $1 \leq i \leq n$,

$$\mathbb{P}(\mathbf{H}_i \mid D) = \frac{\mathbb{P}(D \mid \mathbf{H}_i)\mathbb{P}(\mathbf{H}_i)}{\sum_{j=1}^{n} \mathbb{P}(D \mid \mathbf{H}_j)\mathbb{P}(\mathbf{H}_j)} . \tag{10.3}$$

PROOF. By the Theorem of Total Probabilities,

$$\sum_{j=1}^{n} \mathbb{P}(D \mid \mathbf{H}_j)\mathbb{P}(\mathbf{H}_j) = \mathbb{P}(D).$$

Thus, (10.3) is equivalent to

$$\mathbb{P}(\mathbf{H}_i \mid D)\mathbb{P}(D) = \mathbb{P}(D \mid \mathbf{H}_i)\mathbb{P}(\mathbf{H}_i),$$

which holds because, by the definition of the conditional probability, both sides are equal to $\mathbb{P}(\mathbf{H}_i \cap D)$. $\square$

10.6. Remarks. In the terminology of discussions often associated with Bayes Theorem, the events $\mathbf{H}_i$ are called *hypotheses* and $\mathbb{P}(\mathbf{H}_i)$ is called the *a priori probability* of hypothesis $\mathbf{H}_i$. The conditional probability $\mathbb{P}(\mathbf{H}_i \,|\, D)$ is referred to as the *a posteriori probability* of hypothesis $\mathbf{H}_i$, that is, the probability of $\mathbf{H}_i$ given the evidence provided by the realization of the event D. This terminology is justified by the following interpretation.

Suppose that some experiment has been performed, and some 'data' (the event D) have been collected. A number of alternative 'theories' (the incompatible events $\mathbf{H}_1, \ldots, \mathbf{H}_n$) are considered. An intuitively appealing question is: what are the respective 'probabilitites' of the various theories, given the data? Bayes Theorem suggests that such 'probabilities' might perhaps be recomputed from the a priori probabilities of these hypotheses via the system of equations (10.3) (for $1 \leq i \leq n$) in Theorem 10.5. However, several questions arise. For instance, in what sense can we talk about the 'probability' of a theory ? At first blush, it seems that the quantities $\mathbb{P}(D \,|\, \mathbf{H}_i)$ should not create difficulties: One might reasonably suppose that any genuine theory $\mathbf{H}_i$ should specify the probability of any conceivable data D that could result from the experiment. But what meaning could be given to the quantities $\mathbb{P}(\mathbf{H}_i \,|\, D)$ or $\mathbb{P}(\mathbf{H}_i)$? More generally, what would be an appropriate sample space in which theories would be events? A concrete example will be helpful.

The Dishonest Gambler

10.7. Example. We go back once again to the die-tossing experiment of Example 9.1, with sample space

$$\Omega = \{1,2,3,4,5,6\} \times \{1,2,3,4,5,6\}. \tag{10.4}$$

A gambler has two dice, one fair, and one loaded. For some reason, he does not know which of the two dice is now in his pocket (just before an important game). More precisely, in the notation of this chapter, we have the two hypotheses or 'theories':

$\mathbf{H}_1$: the die in his pocket is fair;

$\mathbf{H}_2$: the die is loaded; in repeated tosses, the relative frequency of a '1' will approximate $\frac{3}{8}$ and the other faces will occur with an equal frequency of about $\frac{1}{8}$.

In the framework that we have developed so far, it makes sense to talk about the 'probability of a theory' only if this theory has a representation as an event in a probability space. Our first step will thus be to enlarge the sample space Ω defined by (10.4) in order to incorporate the two theories $\mathbf{H}_1$ and $\mathbf{H}_2$. There are several correct ways of constructing this enlarged sample space. We only consider one of them here, which seems to be the most natural one.

Consider the sample space

$$\Omega' = \{h_1, h_2\} \times \Omega$$

in which h_1 and h_2 stand for the two hypotheses. (The correspondence with $\mathbf{H}_1$ and $\mathbf{H}_2$, which will be events in the new probability space, is made clear below.) For the field of sets on Ω', we simply take $2^{\Omega'}$, the set of all subsets of Ω'. The two hypotheses are now identified with the two events:

$$\mathbf{H}_i = \{(h_i, j, k) \mid j, k \in \{1, 2, 3, 4, 5, 6\}\}, \qquad i = 1, 2.$$

We have thus:

$$\Omega' = \mathbf{H}_1 \cup \mathbf{H}_2, \qquad \mathbf{H}_1 \cap \mathbf{H}_2 = \emptyset.$$

Finally, we need to construct a probability measure $\mathbb{P}$. Notice that there are many aspects of such a probability measure that are already fixed. For instance, $\mathbb{P}$ must satisfy

$$\mathbb{P}\left(\{(h_1, 1, 1)\} \mid \mathbf{H}_1\right) = \frac{1}{6} \times \frac{1}{6}, \qquad (10.5)$$

$$\mathbb{P}\left(\{(h_2, 1, 1)\} \mid \mathbf{H}_2\right) = \frac{3}{8} \times \frac{3}{8}. \qquad (10.6)$$

Indeed, $\{(h_1, 1, 1)\}$ represents the compound event '$\mathbf{H}_1$ and getting a '1' on both trials' in our new sample space and we reasonably assume that, under (that is, given) either of the two hypotheses, the events 'a '1' on the first trial' and 'a '1' on the second trial' are independent. Under the condition that $\mathbf{H}_1$ is true (for example, the die is fair), the probability of this event is $(\frac{1}{6})^2$, as stated in Eq. (10.5). On the other hand, if $\mathbf{H}_2$ is true, then the die is loaded; so, the conditional probability of getting a '1' on both trials given the event $\mathbf{H}_2$ is $(\frac{3}{8})^2$ as in Eq. (10.6). In the same vein, we must have for any $k, \ell \in \{1, 2, 3, 4, 5, 6\}$

$$\mathbb{P}\left(\{(h_i, k, \ell)\} \mid \mathbf{H}_j\right) = 0 \qquad \text{if } i \neq j$$

and also

$$\mathbb{P}\left(\{(h_1, j, \ell)\} \mid \mathbf{H}_1\right) = \left(\frac{1}{6}\right)^2, \quad \mathbb{P}\left(\{(h_2, 1, 2)\} \mid \mathbf{H}_2\right) = \frac{3}{8} \times \frac{1}{8},$$

$$\mathbb{P}\left(\{(h_2, 2, 3)\} \mid \mathbf{H}_2\right) = \left(\frac{1}{8}\right)^2, \quad \text{etc.}$$

In general, for every point $x \in \Omega'$, we know the value of the conditional probability $\mathbb{P}(\{x\} \mid \mathbf{H}_i)$, for $i = 1, 2$. This information, however, is not sufficient to determine the probability measure $\mathbb{P}$. Some critical information is missing: If we knew the a priori probabilities $\mathbb{P}(\mathbf{H}_i)$ of the two hypotheses, then $\mathbb{P}\{x\}$ would be specified for all $x \in \Omega'$ by the equation

$$\mathbb{P}(\{x\}) = \mathbb{P}(\{x\} \mid \mathbf{H}_1)\mathbb{P}(\mathbf{H}_1) + \mathbb{P}(\{x\} \mid \mathbf{H}_2)\mathbb{P}(\mathbf{H}_2).$$

In other words, if we knew the values of the $\mathbb{P}(\mathbf{H}_i)$'s, the probability measure $\mathbb{P}$ would be determined from Theorem 4.6. (The argument here is that the $\mathbb{P}(\{x\})$ values define a probability distribution on Ω'.) The key question is thus: How do we assign values to the a priori probabilities of the two theories $\mathbf{H}_1$ and $\mathbf{H}_2$? A position frequently adopted by the statistician is based on the so-called *Principle of Indifference*: If there are no reasons to prefer $\mathbf{H}_1$ over $\mathbf{H}_2$, we set

$$\mathbb{P}(\mathbf{H}_1) = \mathbb{P}(\mathbf{H}_2) = \frac{1}{2}. \tag{10.7}$$

In some situations, this will not be satisfactory. For instance, the gambler may vaguely remember having left one of the dice, say the the loaded one, in a drawer in his hotel room. In such a case, the gambler might then be more willing to accept

$$\mathbb{P}(\mathbf{H}_1) = .7, \qquad \mathbb{P}(\mathbf{H}_2) = .3 \tag{10.8}$$

rather than (10.7), for the values of the a priori probabilities.

In any event, from (10.7), (10.8), or any other any such assignment, we can define the probability measure $\mathbb{P}$ on $2^{\Omega'}$ by

$$\mathbb{P}(A) = \sum_{x \in A} \mathbb{P}(\{x\} \mid \mathbf{H}_1)\mathbb{P}(\mathbf{H}_1) + \sum_{x \in A} \mathbb{P}(\{x\} \mid \mathbf{H}_2)\mathbb{P}(\mathbf{H}_2),$$

for all $A \subseteq \Omega'$. For the remainder of this discussion, we shall take the viewpoint of the Principle of Indifference, and assume that (10.7) holds.

Let us compute the a posteriori probabilities of the two theories $\mathbf{H}_1$ and $\mathbf{H}_2$, given the data 'a '1' on both trials,' which corresponds to the event

$$D = \{(H_1, 1, 1), (H_2, 1, 1)\} \subseteq \Omega.'$$

Using Bayes Theorem, we have

$$\mathbb{P}(\mathbf{H}_1 \mid D) = \frac{\mathbb{P}(D \mid \mathbf{H}_1)\mathbb{P}(\mathbf{H}_1)}{\mathbb{P}(D \mid \mathbf{H}_1)\mathbb{P}(\mathbf{H}_1) + \mathbb{P}(D \mid \mathbf{H}_2)\mathbb{P}(\mathbf{H}_2)}$$

$$= \frac{(1/6)^2(1/2)}{(1/6)^2(1/2) + (3/8)^2(1/2)} = \frac{16}{97}.$$

Since $\mathbb{P}(\mathbf{H}_2 \mid D) = 1 - \mathbb{P}(\mathbf{H}_1 \mid D)$ because $\mathbf{H}_2 = \overline{\mathbf{H}}_1$, we also have

$$\mathbb{P}(\mathbf{H}_2 \mid D) = \frac{81}{97}.$$

Thus, on the basis of the data D, it would be reasonable for an experimenter accepting the Principle of Indifference to prefer $\mathbf{H}_2$ over $\mathbf{H}_1$.

This type of argument can be generalized to any arbitrary finite collection $(\mathbf{H}_i)_{1 \leq i \leq n}$ of theories. For any data D and any two theories $\mathbf{H}_i, \mathbf{H}_j$, we obtain (assuming that the denominators do not vanish), by Bayes Theorem and the Principle of Indifference

$$\frac{\mathbb{P}(\mathbf{H}_i \mid D)}{\mathbb{P}(\mathbf{H}_j \mid D)} = \frac{\mathbb{P}(D \mid \mathbf{H}_i)\,\mathbb{P}(\mathbf{H}_i)}{\mathbb{P}(D \mid \mathbf{H}_j)\,\mathbb{P}(\mathbf{H}_j)} = \frac{\mathbb{P}(D \mid \mathbf{H}_i)}{\mathbb{P}(D \mid \mathbf{H}_j)} \qquad (10.9)$$

and so

$$\frac{\mathbb{P}(\mathbf{H}_i \mid D)}{\mathbb{P}(\mathbf{H}_j \mid D)} = \frac{\mathbb{P}(D \mid \mathbf{H}_i)}{\mathbb{P}(D \mid \mathbf{H}_j)}. \qquad (10.10)$$

THIS IMPLIES THAT, UNDER THE PRINCIPLE OF INDIFFERENCE, THE THEORY HAVING THE MAXIMUM PROBABILITY GIVEN THE DATA IS ALSO THE THEORY MAXIMIZING THE LIKELIHOOD (OR THE PROBABILITY) OF THE DATA.

This remark introduces and motivates the concept of 'maximum likelihood' which is the topic of the next chapter. Note that in practice, one never bothers explicitly to introduce the enlarged sample space within which the competing theories are bona fide events. The researcher proceeds with the type of calculation exemplified by (10.9) and (10.10) as if such a sample space had been constructed.

Problems

1. Let A and B be two independent events in some probability space, with probabilities $\mathbb{P}(A) = x$ and $\mathbb{P}(B) = y$. We suppose that $0 < x < 1$ and $0 < y < 1$. Give an expression, in terms of the quantities x and y, for the probabilities of the following events:

   (a) neither the event A nor the event B occurs;

   (b) A occurs but B does not occur;

   (c) either A does not occur or B does not occur;

   (d) either A occurs or B does not occur;

   (e) either A but not B occurs or B but not A occurs;

   (f) A but not B occurs; however, B but not A fails to occur.

2. If A and B are independent events, what can be said about $\bar{A}$ and $\bar{B}$?

3. Show that events A and B are independent if $\mathbb{P}(B|A) = \mathbb{P}(B|\bar{A})$. Does the converse also hold?

4. For each of the following, state whether the two events are independent:

   (a) in three tosses of a fair coin, the events 'at least two tosses are heads' and 'at least one toss is a tail';

   (b) in a roll of two fair dice, the events 'the sum of the dice is odd' and 'the second die is a multiple of 3';

   (c) in choosing two numbers at random from the interval $]0, 1[$, the events 'the sum of the numbers is greater than 1' and 'the second number is greater than the first.'

5. Archie and Nero decide to play the following game of dice. Archie always plays first and tosses the die. Suppose that Archie's result is i. Then, Nero picks the die and plays. If his result is smaller than i, he is supposed to continue playing, until he obtains a result $j \geq i$. If $i = j$, Nero loses \$2, which he gives to Archie. Otherwise, he wins and receives \$$j - i$. You may assume that the die is fair.

   (a) Construct a sample space for this situation.

   (b) Construct the appropriate probability distribution.

   (c) Compute the probability that Nero will: (a) win; (b) win \$2.

   (d) Are the events: 'Archie's result is even' and 'Nero's result is odd' independent?

   (e) Which player would you rather be? Justify.

6. Consider the two competing theories discussed in Example 10.7. Which of $\mathbf{H}_1$ or $\mathbf{H}_2$ should be favored under the a priori probabilities given by (10.8), supposing that, in the course of three trials, two 1's and one 6 were observed? (Note that there is an ambiguity in this statement. Think of two reasonable interpretations and compute the corresponding probabilities of the hypotheses.)

7. The definition of independence is extended to three or more events as follows: Events $A_1, A_2, \ldots, A_n$ are *independent* if for all subsequences $A_{1'}, A_{2'}, \ldots, A_{r'}$ of $A_1, A_2, \ldots, A_n$,

$$\mathbb{P}(A_{1'} \cap A_{2'} \cap \cdots \cap A_{r'}) = \mathbb{P}(A_{1'})\mathbb{P}(A_{2'}) \cdots \mathbb{P}(A_{r'}). \qquad (10.11)$$

It may be tempting to suppose that Eq.(10.11) holds if we only assume that the events A_i are pairwise independent, that is, $\mathbb{P}(A_i \cap A_j) = \mathbb{P}(A_i)\mathbb{P}(A_j)$ for all $i \neq j$. Devise an example to show that this is not the case.

8. A university has twice as many undergraduates as graduates, with 25% of the graduate population living on campus and 10% of the undergraduate population living on campus. If a student living on campus is chosen at random, what is the probability that this person is an undergraduate?

9. A certain HIV test detects the virus in 98% of infected subjects. Unfortunately, the test also classifies as infected about 0.8% of all noninfected subjects. If 0.5% of the population is infected with HIV, what is the probability that a person who tests positive actually has the virus?

10. Archie is into traveling again. This time, he is stranded in a war zone in the faraway land of the Xhosas. His only chance of survival is to make it to Sudland, which is 100 miles away, and can be reached either by crossing a desert or by going through the mountains. Both are very chancy enterprises. Only 38% of the people who try to make it to Sudland succeed. Archie knows that about $\frac{4}{5}$ who attempt to reach Sudland try to cross the desert. It is also known that the probability of reaching Sudland by the mountain route is 90%. Archie wants to know what his chances of survival are if he goes through the desert. Can you help him?

11. Archie has to go to school this morning for an important probability exam. Unfortunately, he wakes up rather late. He could either take the bus or use the family car. However, this bus is often late in arriving to school. (About 25% of the time, but Archie does not know that for sure.) On the other hand, for a variety of reasons, the family car is quite unreliable (gasoline tank empty, battery completely discharged, etc.). Archie's estimate (based on a long series of experiments) is that were he to use the family car, he would make in on time only with probability .6. Another difficulty is that if he tries the family car, he would certainly miss the bus. Unable to make a rational choice, Archie throws an unbiased coin to decide his mode of transportation. You know that Archie made it on time to school for the exam. What is the probability that he has in fact used the family car?

Chapter 11

The Principle of Maximum Likelihood

For a brief incursion into statistics, we now turn to the important 'Principle of Maximum Likelihood.' According to this principle, to which the name of R.A. Fisher is attached, the researcher should favor, among all the competing theories in a specific class, that which maximizes the likelihood of the data. This principle was motivated in Chapter 10 by the following observation. Suppose that we have a finite collection $(\mathbf{H}_i)_{1 \leq i \leq n}$ of events, each of which can be regarded as representing one of n theories competing for the explanation[1] of some data D. Applying both Bayes Theorem and the Principle of Indifference—which states that all the competing theories are a priori equally probable—we obtained the equation

$$\frac{\mathbb{P}(\mathbf{H}_i \mid D)}{\mathbb{P}(\mathbf{H}_j \mid D)} = \frac{\mathbb{P}(D \mid \mathbf{H}_i)}{\mathbb{P}(D \mid \mathbf{H}_j)}. \tag{11.1}$$

One consequence of Eq. (11.1) is that the theory which is the most likely for a given set of data D is also that theory making the data D the most likely. Notwithstanding Eq. (11.1), one does not usually bother with constructing explicitly a sample space containing the competing theories as events. More simply, one considers, for a fixed sample space Ω with field of sets $\mathcal{F}$, a class $\mathcal{C}$ of theories each of which induces its own probability measure on $\mathcal{F}$. For example, the theories in the class $\mathcal{C}$ might differ from each other by the values

---

[1] Or 'prediction,' in the standard lingo of mathematical modeling.

of one or more parameters. The best theory in the class $\mathcal{C}$ for a given set of data, in the sense of the Principle of Maximum Likelihood, is the theory whose probability measure gives the data the maximum chance of realization.

The applications of this principle are very general. In particular, they are not limited to probability spaces that are only finitely additive. Also, the number of competing theories does not have to be finite. On the other hand, the class of competing theories in consideration should be homogeneous. In most cases, the Principle of Maximum Likelihood is applied to a class of theories whose members only differ by one or more parameters.

So far, these ideas have only been discussed in the abstract. The purpose of this chapter is to examine two examples in detail. In both of them, the class of competing theories form a continuum.

A Continuum of Theories

11.1. Data and Theory. Suppose that the data of N consecutive coin tosses have been collected. An appropriate sample space is thus

$$\Omega = \underbrace{\{H,T\} \times \{H,T\} \times \ldots \times \{H,T\}}_{N \text{ factors}},$$

with 2^{Ω} as the field of sets. We do not know which coin has been used. For some reason, we want to consider all possible biases. We assume that the probability of a head occurring on one toss is some number θ, with $0 < \theta < 1$ independent of the trial number. For any particular value of θ, we define as follows a probability distribution on Ω. Let x be some arbitrary point in Ω, say

$$x = (H,T,T,\ldots,T,H,T).$$

We assume that the successive trials are independent (that is, that the events 'head on first trial,' 'tail on second trial,' etc. are independent), and we set the probability of the point x to be

$$p_\theta(x) = \theta(1-\theta)(1-\theta)\ldots(1-\theta)\theta(1-\theta),$$

since the probability of a head on the first trial is θ, the probability of a tail on the second trial is $1-\theta$, etc. In general, we define

$$p_\theta(x) = \theta^n(1-\theta)^{N-n}$$

for any point $x \in \Omega$ in which the N trials involve exactly n heads.

It is not difficult to check that, for each value of θ, the function $p_\theta : \Omega \to [0, 1]$ is a probability distribution on Ω (that is, $p_\theta(x) \geq 0$ for all $x \in \Omega$, and $\sum_{x \in \Omega} p_\theta(x) = 1$; cf. Problem 5). We know from Theorem 4.6 that p_θ may be used to define a probability measure $\mathbb{P}_\theta$, satisfying $\mathbb{P}_\theta(\{x\}) = p_\theta(x)$, for all $x \in \Omega$. Suppose that such a construction has been made. We have thus a collection of probability spaces $(\Omega, \mathcal{F}, \mathbb{P}_\theta)$. Each of these probability spaces corresponds to a possible theory. In this framework, θ is called a *parameter* and $\{\theta \mid 0 < \theta < 1\}$ the *parameter space*. Let

$$D = \{(H, T, T, \ldots, T, H, T)\}$$

be the event in $\mathcal{F}$ representing the data collected in some experiment. Suppose, in particular, that D involves n heads and $N - n$ tails.

11.2. Maximum Likelihood Estimate of θ. What is the 'best' theory for D? In other words, what is the 'best' value for θ? From the viewpoint of the Principle of Maximum Likelihood, we should answer these questions by searching for the value of θ which renders maximum the likehood of the data, that is, the quantity

$$\mathbb{P}_\theta(D) = \theta^n (1 - \theta)^{N-n}. \tag{11.2}$$

Observe that some value of θ maximizes $\mathbb{P}_\theta(D)$ if and only if it also maximizes

$$\log \mathbb{P}_\theta(D) = n \log \theta + (N - n) \log(1 - \theta). \tag{11.3}$$

Taking derivatives with respect to θ in (11.3), equating the result to zero, and solving for θ gives

$$\theta = \frac{n}{N}.$$

This result amounts to estimating θ by the relative frequency of heads in the N trials, which is a traditional procedure.

In the context of the Principle of Maximum Likelihood, the quantity $\frac{n}{N}$ is called a *maximum likelihood estimate* of the parameter θ. Maximum likelihood estimators are among the most useful and best understood in statistical theory. In this example, one could have guessed that $\frac{n}{N}$ was a good estimator of θ. Such a guess is much more difficult to venture in our next example.

Three-Trial Learning

11.3. Data and Theory. Consider a three-trial learning (or problem solving) experiment which is so constrained that the subject cannot give the correct response by chance. (For example, the number of possible responses is very large.) On each trial, the experimenter records whether the response is correct or not, and informs the subject of this result by indicating the correct response. The following 'insight' theory is assumed. We shall describe the theory informally, and also give a representation in the form of a graph (see Fig. 11.1).

There are two possible cognitive states for the subject, labeled $\overline{K}$ and K. A subject in the naive state $\overline{K}$ has not yet discovered the rule governing the correct responses. The state K represents the case in which the subject knows the rule. The subject starts the experiment in state $\overline{K}$. On each trial, there is a probability a that the subject discovers the rule, that is, moves from state $\overline{K}$ to state K. The subject never forgets the rule: When in state K on some trial, the subject always remains in state K for the next trial. As mentioned earlier, the probability that the subject gives the correct response when in state $\overline{K}$ is equal to zero. We also assume that the probability of a correct response when in state K is equal to one. (This assumption is dropped in Problem 3.)

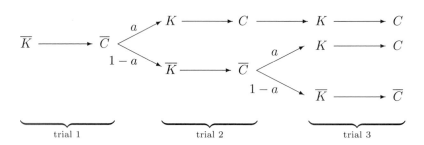

Figure 11.1: Transition diagram in the 3-trial learning experiment.

We denote by C and $\overline{C}$ the correct and incorrect responses, respectively. By convention, the experiment lasts only three trials. We can thus take

$$\Omega = \left(\{C, \overline{C}\} \times \{K, \overline{K}\} \right)^3$$

as our sample space, with 2^Ω as our field of events. The graph of

Fig. 11.1 represents the possible transitions between trials.

It is not difficult to see that the diagram of Fig. 11.1 specifies a probability measure $\mathbb{P}$ on all subsets of Ω, up to the value of one parameter a. According to this theory, there are only three possible cases for an observed triple of successive responses, namely,

1. $(\overline{C}, C, C)$;

2. $(\overline{C}, \overline{C}, C)$;

3. $(\overline{C}, \overline{C}, \overline{C})$.

All other triples of responses are assigned zero probability. (The student should verify this.) Suppose that, in an experiment involving N subjects, these three cases occur with frequencies N_1, N_2, and N_3, respectively. Assuming that all the subjects are governed by the same value[2] of the parameter a, the likelihood of these data is easily computed to be

$$L(a; N_1, N_2, N_3) = a^{N_1} \left((1-a)a\right)^{N_2} \left((1-a)^2\right)^{N_3}. \tag{11.4}$$

11.4. Maximum Likelihood Estimate of a. The maximum likelihood estimate of a can be obtained by maximizing for the variable a the function L in (11.4) for $0 < a \leq 1$. In other words, the expression in (11.4) is regarded as a function of the parameter a, and the maximum of that function, for a varying in the interval $]0, 1]$, is determined. This can be achieved by standard methods of calculus. Taking logarithms in (11.4), taking derivatives with respect to a, equating the result to zero, and solving for a, yields the estimate

$$\widehat{a} = \frac{N_1 + N_2}{N_1 + 2N_2 + 2N_3}. \tag{11.5}$$

11.5. Remark. In this theory, the cognitive states K and $\overline{K}$ are observable because only correct responses are given in state K, and incorrect responses in state $\overline{K}$. In Problem 3, however, we ask the student to generalize the theory by allowing careless errors in state K. The states K and $\overline{K}$ become unobservable (in trials 2 and 3). Despite its simplicity, this generalized theory is remarkable

---

[2]A highly questionable assumption for human subjects, but one that experimental psychologists accept with abandon for rats.

in that the points and events of the sample space involve unobservable entities. This illustrates a device frequently used in the application of scientific theories: A scientific theory sets constraints on the possible data via some unobservable entities, which on the one hand provide an explanation of the data, and on the other hand are assessed or evaluated by the data. A celebrated example is Mendelean genetics before 1915.[3]

Problems

1. Verify in detail that the only events of the sample space which have a nonzero probability are those corresponding to the three triples of responses $(\bar{C}, C, C), (\bar{C}, \bar{C}, C)$, and $(\bar{C}, \bar{C}, \bar{C})$.

2. Verify that Eq. (11.5) gives a correct maximum likelihood estimate of the parameter a.

3. Consider a generalization of the theory described in 11.3 in which a subject in state K can make careless errors. Denote by θ the probability of a careless error.

   (a) Give a correct and complete statement of the theory.

   (b) Compute, in terms of the parameters θ and a, the probabilities of all the observable triples of responses.

   (c) Examine possible ways of estimating the parameters.

4. Consider now a generalization of the theory in 11.3 in which the parameter a is allowed to vary with the trial. Specifically, suppose that the probability the subject discovers the rule immediately after the first trial is a_1 and that the probability he discovers the rule immediately after the second trial is a_2.

   (a) Compute, in terms of the parameters a_1 and a_2, the probabilities of all the observable triples of responses.

   (b) Find the maximum likelihood estimates of a_1 and a_2 in terms of the frequencies with which these triples occur in an experiment involving N subjects. What do you notice about these estimates?

5. For the coin-tossing experiment in 11.1, show that, for any $0 \leq \theta \leq 1$, the function $p_\theta : \Omega \to [0, 1]$ defined by $p_\theta(x) = \theta^n (1 - \theta)^{N-n}$ (for any N-tuple $x = (\ldots, H, \ldots, T, \ldots)$ having exactly n terms H) is a probability distribution on Ω.

---

[3]That is, before the publication of the genetic map of the chromosomes of *Drosophila melanogaster* by Morgan, Sturtevant, Muller, and Bridges.

Chapter 12

Random Variables

There are circumstances in which the experimenter is only interested in some numerical aspects of the possible outcomes. In such a case, it is convenient to set up special numerical functions defined on the sample space and describing these aspects.

A Real Function on the Sample Space

12.1. Example. We go back once more to the experiment considered in 9.1 and 10.7, which involved two consecutive tosses of a fair die. We adopted as the sample space

$$\Omega = \{(1,1), (1,2), \ldots, (2,1), \ldots, (6,6)\}, \qquad (12.1)$$

with a probability measure $\mathbb{P}$ defined on all subsets of Ω by the probability distribution

$$p(\omega) = \frac{1}{36}, \qquad \text{for all } \omega \in \Omega.$$

Suppose, however, that we are only concerned with the sum $j + k$ of the two numbers obtained in an outcome (j, k). The relevant information can be obtained from the transformation

$$(j, k) \longmapsto j + k \qquad ((j,k) \in \Omega) \qquad (12.2)$$

recoding each outcome into the sum of its terms. Such a recoding implicitly defines a new sample space

$$\Omega' = \{2, 3 \ldots, 12\} \subseteq \mathbb{R}$$

with probability distribution

$$p'(2) = p(1,1) = \frac{1}{36},$$

$$p'(3) = p(1,2) + p(2,1) = \frac{2}{36},$$

$$\cdots$$

$$p'(12) = p(6,6) = \frac{1}{36}.$$

This example illustrates a fundamental concept in probability theory. The recoding represented in (12.2) defines a function, which we denote by $\mathbf{Y}$, mapping Ω into $\mathbb{R}$, with

$$\mathbf{Y}(j,k) = j + k \qquad\qquad ((i,j) \in \Omega). \qquad (12.3)$$

Such a function is usually called a 'random variable.' An exact definition of this concept will be given shortly. For the moment, it suffices to remember that a random variable is a function mapping the sample space into the set $\mathbb{R}$ of all real numbers. Note in passing the customary usage of denoting random variables by bold capital letters. Let us consider other examples.

Further Examples and a Definition

12.2. Example. Several random variables can be defined on the same sample space, focusing on different aspects of the outcomes. In the sample space Ω specified by (12.1), suppose that we only consider the result of the first toss. We can define a random variable $\mathbf{Y}_1$ selecting that information by the equation

$$\mathbf{Y}_1(j,k) = j. \qquad (12.4)$$

Obviously, a random variable $\mathbf{Y}_2$ can be similarly defined to retain the result of the second toss. A complete formalization of this die tossing experiment is thus obtained from the pair of random variables $(\mathbf{Y}_1, \mathbf{Y}_2)$. Notice that we have, with $\mathbf{Y}$ defined as in (12.3),

$$\mathbf{Y}(\omega) = \mathbf{Y}_1(\omega) + \mathbf{Y}_2(\omega), \qquad \forall \omega \in \Omega. \qquad (12.5)$$

The fact that random variables assign numerical values to the outcomes gives access to various quantitative manipulations, such as that illustrated by Eq. (12.5).

12.3. Conventions. While a random variable is formally a function defined on the sample space, the argument of the function is often omitted for simplicity. For example, we would typically write $\mathbf{Y} = \mathbf{Y}_1 + \mathbf{Y}_2$ as a compact expression of (12.5). We also routinely write

$$\mathbb{P}(\mathbf{Y}_1 = 1, \mathbf{Y} \leq 4) = \mathbb{P}\{\omega \in \Omega_1 \mid \mathbf{Y}_1(\omega) = 1 \ \& \ \mathbf{Y}(\omega) \leq 4\}, \quad (12.6)$$

which leads to

$$\mathbb{P}(\mathbf{Y}_1 = 1 \mid \mathbf{Y} \leq 4) = \frac{\mathbb{P}(\mathbf{Y}_1 = 1, \mathbf{Y} \leq 4)}{\mathbb{P}(\mathbf{Y} \leq 4)}$$

$$= \frac{|\{(1,1),(1,2),(1,3)\}|}{|\{(1,1),(1,2),(2,1),(1,3),(3,1),(2,2)\}|} = \frac{1}{2}.$$

Note that the comma in $\mathbb{P}(\mathbf{Y}_1 = 1, \mathbf{Y} \leq 4)$ stands for the logical conjunction '&' in the right-hand side of (12.6). Such writing conventions are standard in probability theory.

12.4. Example. In many cases, a complete description of a random phenomenon is given solely in terms of random variables. In Example 2.3, a rat is learning a T-maze. A typical outcome of this experiment is a sequence $(L, L, R, \ldots, R, \ldots)$ describing the animal's choices on successive trials. Thus, in this particular case, the rat turned left on the first two trials, then right, etc. For every positive integer n, we define a random variable $\mathbf{R}_n$ describing the animal's behavior on that trial. Specifically, for every sample point (or sequence) ω, we set

$$\mathbf{R}_n(\omega) = \begin{cases} 1 & \text{if the } n\text{th term of } \omega \text{ is } L \\ 0 & \text{otherwise.} \end{cases}$$

A complete formal description of the experiment can thus be given in terms of the sequence of random variables $\{\mathbf{R}_n \mid n = 1, 2, \ldots\}$, or more compactly $(\mathbf{R}_n)_{n \in \mathbb{N}}$. Such random variables, which only take the two values 0 and 1, are sometimes called *Bernoulli* random variables (cf. the Bernoulli distribution defined in 8.2).

12.5. Example. The recoding specified by a random variable may obviously involve a loss of information on the outcomes, in the sense that several outcomes sharing some property may be assigned the same value. For instance, in the die tossing experiment of Example 12.1, we had: $\mathbf{Y}[(1,2)] = \mathbf{Y}[(2,1)] = 3$. This feature is by no means necessary. In some important situations, a random variable will be a one-to-one function.

In the experiment of Example 2.2, a coin is tossed until a head appears. The chosen sample space was the set

$$\mathbb{N} = \{1, 2, \ldots, k, \ldots\}.$$

Thus, an outcome k in $\mathbb{N}$ symbolizes the result that the first head appeared on the kth toss. In some cases, it is natural and convenient to define the identity random variable $\mathbf{Z}(k) = k$, for all $k \in \mathbb{N}$. While it may seem that not much is accomplished by such a trivial recoding, it is nevertheless convenient to use random variables in this and other similar cases in order to ensure homogeneity of language and notation.

12.6. Example. Similarly, in the reaction time experiment of Example 2.5, it makes sense—and is actually customary—to use the identity random variable $\mathbf{T} : \mathbb{R} \mapsto \mathbb{R}$, with $\mathbf{T}(t) = t$ for all real numbers t.

12.7. Example. Not all real valued functions on a sample space are random variables. Consider the sample space $\Omega_1 = \{1, 2, 3, 4, 5\}$ with field of events

$$\mathcal{G} = \{\emptyset, \{1, 2\}, \{3, 4, 5\}, \{1, 2, 3, 4, 5\}\},$$

and equipped with the counting measure $\mathbb{P}$ (cf. 4.2); thus, for instance

$$\mathbb{P}\{3, 4, 5\} = \frac{|\{3, 4, 5\}|}{|\Omega_2|} = \frac{3}{5}.$$

Suppose that one is only interested in the fact that the observed outcome is odd or even. In the spirit of the preceding examples, we certainly can define a function $D : \Omega_1 \mapsto \mathbb{R}$ by the formula

$$D(i) = \begin{cases} 1 & \text{if } i \text{ is odd} \\ 0 & \text{otherwise.} \end{cases}$$

Thus, D is a real valued function on the sample space. However, D is not a bona fide random variable. The difficulty is that we cannot assign a probability to a statement such as 'the value taken by D is even,' since the subset of outcomes $\{i \in \Omega_1 \mid i \text{ is even}\} = \{2, 4\}$ is not an event (that is, a member of $\mathcal{G}$), and thus $\mathbb{P}\{2, 4\}$ is not defined.

A general definition of the concept of a random variable is given below.

12.8. Definition. Let $(\Omega, \mathcal{F}, \mathbb{P})$ be a finitely additive probability space. Let $\mathbf{X}$ be a mapping of the sample space into the set $\mathbb{R}$ of all real numbers. Then $\mathbf{X}$ is a *random variable* if, for all $x \in \mathbb{R}$, the set $\{\omega \in \Omega \mid \mathbf{X}(\omega) \le x\}$ is an event, that is, a member of $\mathcal{F}$. In this case, the quantity

$$\mathbb{P}(\mathbf{X} \le x) = \mathbb{P}\{\omega \in \Omega \mid \mathbf{X}(\omega) \le x\} \tag{12.7}$$

is defined for every real number x. (Notice that the left-hand side makes use of the compact notation introduced in 12.3.) According to this definition, the function D of Example 12.7 is not a random variable. For instance, $\{i \in \Omega_1 \mid D(i) \le 0\} = \{2, 4\}$ is not an event.

A random variable is called *finite* or *countable* if its range is finite or countable, respectively. For instance, the random variables $\mathbf{Y}$, $\mathbf{Y_1}$, and $\mathbf{Y_2}$ of 12.2 are finite, while the random variable $\mathbf{Z}$ of 12.5 is countable. A random variable which is finite or countable is called *discrete*. In the examples of this chapter, we shall be mostly concerned with discrete random variables. However, all the definitions are valid in the general case.

Several random variables defined on the same sample space are said to be *jointly distributed* (on that sample space). For example, the random variables $\mathbf{Y}$, $\mathbf{Y_1}$, and $\mathbf{Y_2}$ are jointly distributed on the sample space $\Omega = \{(1, 1), (1, 2), \dots, (2, 1), \dots, (6, 6)\}$ of Examples 12.1 and 12.2.

Results on Random Variables

12.9. Theorem. *Let $\mathbf{X}$ be a random variable in a probability space $(\Omega, \mathcal{F}, \mathbb{P})$ and let x be any real number. Then*

$$\{\omega \in \Omega \mid \mathbf{X}(\omega) > x\} = \overline{\{\omega \in \Omega \mid \mathbf{X}(\omega) \le x\}} \tag{12.8}$$

$$\{\omega \in \Omega \mid \mathbf{X}(\omega) \ge x\} = \cap_{n=1}^{\infty}\{\omega \in \Omega \mid \mathbf{X}(\omega) > x - \frac{1}{n}\} \tag{12.9}$$

$$\{\omega \in \Omega \mid \mathbf{X}(\omega) < x\} = \overline{\{\omega \in \Omega \mid \mathbf{X}(\omega) \ge x\}} \tag{12.10}$$

Consequently, the three formulas on the left represent events.

(We recall that the phrase 'probability space' refers to a σ-field with a countably additive probability measure; cf. 3.4 and 4.1.)

PROOF. The three equations are straightforward results from set theory. The first set $\{\omega \in \Omega \,|\, \mathbf{X}(\omega) > x\}$ is an event because it is the complement of an event (see 3.4). The set in the right hand side of (12.9) is an event because it is the countable intersection of events. Finally, $\{\omega \in \Omega \,|\, \mathbf{X}(\omega) < x\}$ is the complement of an event. $\square$

12.10. Conventions. In practice, it often happens that the sample space is not specified. In probabilistic lingo, the sentence 'Let $\mathbf{W}$ and $\mathbf{V}$ be two random variables' means—if no particular sample space is mentioned—that the random variables are jointly distributed and that the sample space is implicitly assumed to be $\mathbb{R}^2$. In other words, both $\mathbf{W}$ and $\mathbf{V}$ must be taken to be defined on $\mathbb{R}^2$ (equipped with its standard Borel field; cf. 3.5), and are in fact the first and the second projection functions $\mathbf{W}(x, y) = x$ and $\mathbf{V}(x, y) = y$. These functions satisfy the conditions of Def. 12.8 and so are random variables. Note also that we will sometimes use the standard abbreviation 'r.v.' for 'random variable.'

12.11. Definition. Two random variables $\mathbf{W}$ and $\mathbf{V}$ are said to be *independent* if for any two Borel subsets A, B (see 3.5), the events $\{\omega \in \Omega \,|\, \mathbf{W}(\omega) \in A\}$ and $\{\omega \in \Omega \,|\, \mathbf{V}(\omega) \in B\}$ are independent, that is,

$$\mathbb{P}(\mathbf{W} \in A, \mathbf{V} \in B) = \mathbb{P}(\mathbf{W} \in A)\,\mathbb{P}(\mathbf{V} \in B). \qquad (12.11)$$

In particular, we must have

$$\mathbb{P}(\mathbf{W} \leq x, \mathbf{V} \leq y) = \mathbb{P}(\mathbf{W} \leq x)\,\mathbb{P}(\mathbf{V} \leq y), \qquad (12.12)$$

for any two real numbers x and y. It can be shown that the two conditions expressed by Eqs. (12.11) and (12.12) are in fact equivalent. (The proof is beyond the scope of these lectures.)

The two r.v.'s $\mathbf{Y}$ and $\mathbf{Y_1}$ of Example 12.2 are not independent because on the one hand $\mathbb{P}(\mathbf{Y} \leq 2, \mathbf{Y_1} \leq 1) = \mathbb{P}(\mathbf{Y} \leq 2) = \frac{1}{36}$, whereas, on the other hand, $\mathbb{P}(\mathbf{Y} \leq 2)\,\mathbb{P}(\mathbf{Y_1} \leq 1) = \frac{1}{36} \times \frac{1}{6} = \frac{1}{216}$.

If $\mathbf{W}$ and $\mathbf{V}$ are discrete r.v., they are independent if and only if

$$\mathbb{P}(\mathbf{W} = x, \mathbf{V} = y) = \mathbb{P}(\mathbf{W} = x)\,\mathbb{P}(\mathbf{V} = y), \qquad (12.13)$$

for all real numbers x and y (see Problem 9).

12.12. Remark. In 12.7, we gave an example of a real valued function on a sample space which was not a random variable. On the other hand, in a countably additive probability space, practically all the operations of arithmetic and analysis, when applied to random variables, yield random variables. For example, if $\mathbf{X}$ and $\mathbf{Y}$ are random variables and α is a real constant, then the following functions on the sample space are also random variables:

$$|\mathbf{X}|, \quad \alpha\mathbf{X}, \quad \mathbf{X}^{\alpha}, \quad \alpha^{\mathbf{X}}, \quad \mathbf{X}+\mathbf{Y}, \quad \mathbf{X}\cdot\mathbf{Y}.$$

We only prove the first of these facts here.

12.13. Theorem. *If $\mathbf{X}$ is a random variable in a probability space $(\Omega, \mathcal{F}, \mathbb{P})$, then $|\mathbf{X}|$ is also a random variable.*

PROOF. Let Ω be the sample space. For every real number x,

$$\{\omega \in \Omega \mid |\mathbf{X}(\omega)| \leq x\} = \{\omega \in \Omega \mid -x \leq \mathbf{X}(\omega) \leq x\}$$
$$= \{\omega \in \Omega \mid \mathbf{X}(\omega) \leq x\} \cap \{\omega \in \Omega \mid -x \leq \mathbf{X}(\omega)\}.$$

By Theorem 12.9, the two sets in the intersection are events. This means that their intersection $\{\omega \in \Omega \mid |\mathbf{X}(\omega)| \leq x\}$ is also an event and the result follows. □

We also formulate, without proof, a more general result:

12.14. Theorem. *Let $\mathbf{X}$ and $\mathbf{Y}$ be two jointly distributed random variables, and let h be a real valued, continuous function on $\mathbb{R}^2$; define*

$$\mathbf{W}(\omega) = h[\mathbf{X}(\omega), \mathbf{Y}(\omega)]$$

for every sample point ω. Then $\mathbf{W}$ is a random variable.

In particular, as mentioned in 12.12, $\mathbf{X}+\mathbf{Y}$ and $\mathbf{X}\cdot\mathbf{Y}$ are random variables. Note that this Theorem implies that if g is a continuous real valued function on $\mathbb{R}$, then $g(\mathbf{X})$ is a random variable. (Take $\mathbf{X} = \mathbf{Y}$ in Theorem 12.14 and define $g[\mathbf{X}(\omega)] = h[\mathbf{X}(\omega), \mathbf{X}(\omega)]$.)

We also omit the proof of the following useful result.

12.15. Theorem. *Let g and f be two real valued, continuous functions on $\mathbb{R}$ and let $\mathbf{X}$ and $\mathbf{Y}$ be two independent random variables. Then the random variables $g(\mathbf{X})$ and $h(\mathbf{Y})$ are also independent.*

12.16. Remark. Two particular cases of the function h in Theorem 12.14 must be pointed out. Let $\mathbf{X}$ and $\mathbf{Y}$ be two jointly distributed random variables, and consider the two functions

$$\mathbf{V} = \max\{\mathbf{X}, \mathbf{Y}\} \quad \text{and} \quad \mathbf{W} = \min\{\mathbf{X}, \mathbf{Y}\}.$$

It is easily verified that the max and the min functions are continuous. Thus, by Theorem 12.14, both $\mathbf{V}$ and $\mathbf{W}$ are random variables. Notice in passing that we have, for any real number x,

$$
\begin{aligned}
\mathbb{P}(\mathbf{V} \leq x) &= \mathbb{P}(\mathbf{X} \leq x, \mathbf{Y} \leq x) \\
&= \mathbb{P}(\mathbf{X} \leq x) + \mathbb{P}(\mathbf{Y} \leq x) - \mathbb{P}(\mathbf{X} \leq x \text{ or } \mathbf{Y} \leq x) \\
&= \mathbb{P}(\mathbf{X} \leq x) + \mathbb{P}(\mathbf{Y} \leq x) - \mathbb{P}(\min\{\mathbf{X}, \mathbf{Y}\} \leq x) \\
&= \mathbb{P}(\mathbf{X} \leq x) + \mathbb{P}(\mathbf{Y} \leq x) - \mathbb{P}(\mathbf{W} \leq x),
\end{aligned}
$$

so that

$$\mathbb{P}(\mathbf{V} \leq x) + \mathbb{P}(\mathbf{W} \leq x) = \mathbb{P}(\mathbf{X} \leq x) + \mathbb{P}(\mathbf{Y} \leq x).$$

Also, if the random variables $\mathbf{X}$ and $\mathbf{Y}$ are independent (cf. 12.11), then

$$\mathbb{P}(\max\{\mathbf{X}, \mathbf{Y}\} \leq x) = \mathbb{P}(\mathbf{X} \leq x)\mathbb{P}(\mathbf{Y} \leq x).$$

Problems

1. Archie tosses a fair coin until the coin comes up heads. Then Nero does the same. Let $\mathbf{T}_1$ be the number of the toss on which Archie's coin first comes up heads, and let $\mathbf{T}_2$ be the number of the toss on which Nero's coin first comes up heads. (a) Are $\mathbf{T}_1$ and $\mathbf{T}_2$ discrete? (b) Are $\mathbf{T}_1$ and $\mathbf{T}_2$ independent? (c) What is $\mathbb{P}(\mathbf{T}_1 + \mathbf{T}_2 \leq 3)$? (d) What is $\mathbb{P}(\mathbf{T}_1 = \mathbf{T}_2)$? (e) What is $\mathbb{P}(\mathbf{T}_1 > \mathbf{T}_2)$?

2. Suppose that $\mathbf{X}$ is a random variable representing the number of heads obtained in n tosses of a coin that has probability α of coming up heads on a single toss. Write $\mathbf{X}$ in terms of n Bernoulli random variables $\mathbf{X}_i$.

3. With $\mathbf{Y}_1$ and $\mathbf{Y}_2$ as in 12.2, construct the probability distributions of the following random variables, as well as the means of these distributions:

   (a) $\mathbf{Y}_1 - \mathbf{Y}_2$;
   (b) $\max\{\mathbf{Y}_1, \mathbf{Y}_2\}$;
   (c) $\frac{\mathbf{Y}_1}{\mathbf{Y}_2}$;
   (d) $|\mathbf{Y}_1 - \mathbf{Y}_2|$.

4. With $\mathbf{Y}$ and $\mathbf{Y}_1$ as in 12.2, compute $\mathbb{P}(2 < \mathbf{Y} < 5 \mid \mathbf{Y}_1 \geq 2)$.

5. Let $\mathbf{Z}_1, \mathbf{Z}_2, \ldots, \mathbf{Z}_n, \ldots$ be a collection of jointly distributed, independent, Bernoulli random variables with a common parameter α.

    (a) Write an expression for $\mathbb{P}(\sum_{n=1}^{20} \mathbf{Z}_n \leq 7)$ in terms of α.

    (b) Find $\mathbb{P}(\sum_{n=1}^{20} \mathbf{Z}_n \leq 7 \mid \mathbf{Z}_1 + \mathbf{Z}_2 + \mathbf{Z}_3 = 2)$ in terms of α.

    (c) Which is greater: $\mathbb{P}(\sum_{n=1}^{20} \mathbf{Z}_n \leq 7 \mid \mathbf{Z}_1 + \mathbf{Z}_2 + \mathbf{Z}_3 = 2)$ or
$$\mathbb{P}(\mathbf{Z}_1 + \mathbf{Z}_2 + \mathbf{Z}_3 = 2 \mid \sum_{n=1}^{20} \mathbf{Z}_n \leq 7)?$$

    (d) Consider the sequence $(\mathbf{Z}_1, \mathbf{Z}_2, \ldots, \mathbf{Z}_n, \ldots)$, and let $\mathbf{Z}_i$ be the first nonzero term in the sequence. Find the probability that i is odd. What can you say about this probability (no matter the value of α)?

6. In the dice tossing situation of Examples 12.1 and 12.2, let $\mathbf{Y}_1$ and $\mathbf{Y}_2$ represent the results of the two tosses, respectively. Suppose that these two random variables are independent. Are the two random variables $\mathbf{Y}_1 + \mathbf{Y}_2$ and $\mathbf{Y}_1 - \mathbf{Y}_2$ also independent?

7. Let $\mathbf{W}$ be a r. v. with a geometric distribution with parameter $\alpha > 0$. Thus, $\mathbb{P}(\mathbf{W} = n) = \alpha^{n-1}(1 - \alpha)$ for any positive integer n. Compute

    (a) $\mathbb{P}(\mathbf{W} = 8 \mid \mathbf{W} \geq 3)$;

    (b) $\mathbb{P}(\mathbf{W} = n \mid \mathbf{W} \geq k)$ for positive integers n and k, with $n \geq k$;

    (c) $\mathbb{P}(\mathbf{W} \geq 8 \mid \mathbf{W} \geq 3)$;

    (d) $\mathbb{P}(\mathbf{W} \geq n \mid \mathbf{W} \geq k)$ for positive integers n and k, with $n \geq k$.

8. Prove directly (that is, without using Theorem 12.14) that if $\mathbf{X}$ and $\mathbf{Y}$ are jointly distributed random variables, then

    (a) $a\mathbf{X} + b$ is a random variable for any $a, b \in \mathbb{R}$;

    (b) $\mathbf{X} + \mathbf{Y}$ is a random variable;

    (c) $\mathbf{X}^n$ is a random variable for any positive integer n;

    (d) $e^{\alpha \mathbf{X}}$ is a random variable for any $\alpha \in \mathbb{R}$.

9. Prove that, in the discrete case, the condition expressed by Eq. (12.12) implies that expressed by Eq. (12.11).

*10. An interesting and useful case of a random variable is the indicator I_X associated with any event X in the sample space. The student should prove this, that is, that I_X is indeed a random variable.

*11. Let $\mathcal{F}$ be any family of subsets of a finite set U. Assume that all the points of U are equiprobable. Find a probabilistic way of stating that there exists a point of U which is contained in at least half of the sets[1] in $\mathcal{F}$. (Hint: use indicator functions.)

---

[1] A famous conjecture, due to Frankl, is that this property always holds for any family $\mathcal{F} \subseteq 2^U$ which is closed under union.

Chapter 13

Distribution Functions

We have seen in Chapter 12 that for any random variable $\mathbf{X}$ the quantity $\mathbb{P}(\mathbf{X} \leq x)$ is always defined (cf. Def. 12.8). This quantity assigns a probability to the half open interval $]-\infty, x]$, that is, the probability that the random variable takes a value in that interval. It turns out that, if these probabilities are known for all such half open intervals $]-\infty, x]$ (thus, for all real numbers x), then a probability measure is determined for all the Borel subsets of $\mathbb{R}$. In other words, we can, in principle, compute the probability that $\mathbf{X}$ will fall in B, for any Borel set B. We shall not prove this fact here. These Borel subsets are the events of $\mathbb{R}$ regarded as a sample space. Thus any random variable can be regarded as a device inducing a probability measure on the set of all the events of $\mathbb{R}$.

These remarks indicate that the function $x \mapsto \mathbb{P}(\mathbf{X} \leq x)$ contains all the relevant information concerning the random variable $\mathbf{X}$ (considered by itself[1]). This function deserves a name and a special notation.

Basic Concept and Examples

13.1. Definition. Let $\mathbf{X}$ be a random variable on some sample space. We shall call the function $F_{\mathbf{X}} : \mathbb{R} \to [0, 1]$ defined by

$$F_{\mathbf{X}}(x) = \mathbb{P}(\mathbf{X} \leq x)$$

the *distribution function* of the random variable $\mathbf{X}$. It is easily

---

[1]That is, for example, without regard to its relation to other random variables.

shown that such a distribution function satisfies the following three
properties:

[DF1] (MONOTONICITY.) If $x < y$, then $F_{\mathbf{X}}(x) \leq F_{\mathbf{X}}(y)$, for all
$x, y \in \mathbb{R}$;

[DF2] (LIMITS.) $\lim_{x \to \infty} F_{\mathbf{X}}(x) = 1, \qquad \lim_{x \to -\infty} F_{\mathbf{X}}(x) = 0$;

[DF3] (RIGHT CONTINUITY.[2]) $\lim_{s \to x+} F_{\mathbf{X}}(s) = F_{\mathbf{X}}(x)$, for all
$x \in \mathbb{R}$.

Conversely, any real valued function F defined on $\mathbb{R}$ satisfying [DF1],
[DF2] and [DF3] can be regarded as the distribution function of some
random variable $\mathbf{X}$. The construction is simple. We take $\mathbb{R}$ to be
the sample space, with the Borel sets as events. The probability
measure $\mathbb{P}$ is defined on every half open interval $]-\infty, x]$ of $\mathbb{R}$
by $\mathbb{P}]-\infty, x] = F(x)$. By the remark above, we know that the
probability measure is now determined for all the Borel sets of $\mathbb{R}$.
Finally, we define the random variable $\mathbf{X}$ by the identity on $\mathbb{R}$. The
graph of an exemplary distribution function is displayed in Fig. 13.1.

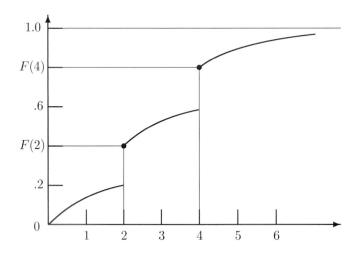

Figure 13.1: Graph of an exemplary distribution function.
We have $\lim_{x \to 2+} F(x) = F(2)$, but $\lim_{x \to 2-} F(x) \neq F(2)$.

---

[2]We recall that a function $f : \mathbb{R} \to \mathbb{R}$ is *right continuous* at some point a if for any $\epsilon > 0$,
there is some $\delta > 0$ such that $\left| f(a) - f(x) \right| < \epsilon$ whenever $a < x < a + \delta$.

Notice that, for any random variable $\mathbf{X}$ with distribution function $F_{\mathbf{X}}$, we have for any two real numbers $x_1 < x_2$

$$\begin{aligned}\mathbb{P}(x_1 < \mathbf{X} \le x_2) &= \mathbb{P}(\mathbf{X} \le x_2) - \mathbb{P}(\mathbf{X} \le x_1)\\ &= F_{\mathbf{X}}(x_2) - F_{\mathbf{X}}(x_1)\end{aligned} \tag{13.1}$$

If moreover $\mathbb{P}\{x_1\} = \mathbb{P}\{x_2\} = 0$, then we also have

$$\begin{aligned}\mathbb{P}(x_1 \le \mathbf{X} \le x_2) &= \mathbb{P}(x_1 \le \mathbf{X} < x_2)\\ &= \mathbb{P}(x_1 < \mathbf{X} < x_2)\\ &= F_{\mathbf{X}}(x_2) - F_{\mathbf{X}}(x_1).\end{aligned} \tag{13.2}$$

These results are useful because, in some important situations, all points of the sample space (or of some interval of it) are assigned probability zero. Examples will be given in Chapter 14. Here, we only consider a few examples of distribution functions in the discrete case.

13.2. Example. In the coin tossing situation of Example 2.1, the field of events is the power set of the sample space

$$\Omega = \{(H, H), (H, T), (T, H), (T, T)\}.$$

Thus, any real valued function on Ω is automatically a random variable. Let us assume that the coin is fair, and let $\mathbf{W}$ be the random variable counting the number of heads, that is, defined by

$$\mathbf{W}[(H, H)] = 2, \quad \mathbf{W}[(H, T)] = \mathbf{W}[(T, H)] = 1, \quad \mathbf{W}[(T, T)] = 0.$$

For any real number $y < 0$, we have

$$\{\omega \in \Omega \,|\, \mathbf{W}(\omega) \le y\} \subseteq \{\omega \in \Omega \,|\, \mathbf{W}(\omega) < 0\} = \emptyset,$$

which yields for the distribution function $F_{\mathbf{W}}$ of $\mathbf{W}$

$$F_{\mathbf{W}}(y) = \mathbb{P}(\mathbf{W} \le y) = \mathbb{P}(\emptyset) = 0, \qquad \text{for } y < 0.$$

For $0 \le y < 1$, we notice that

$$\{(T, T)\} = \{\omega \in \Omega \,|\, \mathbf{W}(\omega) = 0\} = \{\omega \in \Omega \,|\, \mathbf{W}(\omega) < 1\}.$$

(Each of the three formulas denote the same event.) Consequently,

$$F_{\mathbf{W}}(y) = \mathbb{P}(\mathbf{W}(\omega) \le y) = \mathbb{P}\{(T, T)\} = .25, \qquad \text{for } 0 \le y < 1.$$

The other cases are easily checked. Summarizing, we obtain

$$
F_{\mathbf{W}}(y) = \begin{cases} 0 & \text{for} & y < 0 \\ .25 & \text{for} & 0 \le y < 1 \\ .75 & \text{for} & 1 \le y < 2 \\ 1 & \text{for} & 2 \le y. \end{cases}
$$

The function $F_{\mathbf{W}}$ is increasing;[3] has the appropriate limit behavior, and is right continuous. Conditions [DF1], [DF2], and [DF3] are thus satisfied.

13.3. Example. The case of the random variable $\mathbf{Z}$ of Example 2.2, in which a coin is tossed until a head appears, is similar, but slightly more involved. Let us suppose that the coin is fair. We clearly have for any real number z,

$$F_{\mathbf{Z}}(z) = \mathbb{P}(\mathbf{Z} \le z) = \mathbb{P}(\emptyset) = 0, \qquad\qquad \text{if } z < 1.$$

Let us compute a few other cases:

$$F_{\mathbf{Z}}(z) = \mathbb{P}(\mathbf{Z} \le z) = \mathbb{P}(\mathbf{Z} = 1) = .5, \qquad \text{if } 1 \le z < 2;$$
$$F_{\mathbf{Z}}(z) = \mathbb{P}(\mathbf{Z} \le z) = \mathbb{P}(\mathbf{Z} = 1) + \mathbb{P}(\mathbf{Z} = 2)$$
$$= .5 + (.5)^2, \qquad\qquad \text{if } 2 \le z < 3.$$

Using induction, we obtain

$$F_{\mathbf{Z}}(z) = \mathbb{P}(\mathbf{Z} \le z) = .5 + (.5)^2 + \ldots + (.5)^n$$
$$= 1 - (.5)^n \qquad\qquad \text{if } n \le z < n+1.$$

Summing up, we get in the general case

$$
F_{\mathbf{Z}}(z) = \begin{cases} 0 & \text{if } z < 1, \\ 1 - (.5)^n & \text{if } 1 \le n \le z < n+1 \text{ for some } n \in \mathbb{N}. \end{cases}
$$

A (partial) graph of $F_{\mathbf{Z}}$ is displayed in Fig. 13.2. As suggested by these computations, there is a probability distribution on $\mathbb{N}$ which is intimately related to the distribution function $F_{\mathbf{Z}}$ and can be constructed from it.

---

[3]But not strictly increasing.

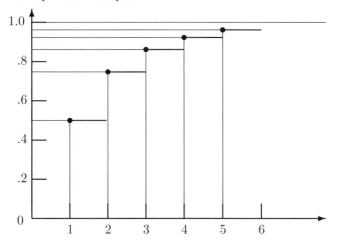

Figure 13.2: Partial graph of the distribution function $F_{\mathbf{Z}}$ of Example 13.3.

Define

$$p(1) = F_{\mathbf{Z}}(1) = .5$$
$$p(2) = F_{\mathbf{Z}}(2) - F_{\mathbf{Z}}(1) = (.5)^2$$
$$\vdots \qquad \vdots$$
$$p(n) = F_{\mathbf{Z}}(n) - F_{\mathbf{Z}}(n-1) = (.5)^n$$

for any natural number n. In fact, p is the geometric distribution with parameter $\alpha = .5$ (see 8.9).

Conversely, the distribution function $F_{\mathbf{Z}}$ can be defined from the geometric distribution p by the formula

$$F_{\mathbf{Z}}(z) = \begin{cases} 0 & \text{for} & z < 1 \\ \sum_{i=1}^{n} p(i) & \text{for} & n \leq z < n+1,\, n = 1, 2 \dots \end{cases} \qquad (13.3)$$

This example illustrates a tight relationship between a distribution function and an associated probability distribution defined on a finite or countable subset of the reals. More generally, we have the following result, the proof of which is left to the student:

13.4. Theorem. *Let $(\Omega, \mathcal{F}, \mathbb{P})$ be a finitely additive probability space. Let $\mathbf{X}$ be a discrete random variable with range $x_1 < x_2 < \cdots < x_i < \dots$ and distribution function $F_{\mathbf{X}}(x) = \mathbb{P}(\mathbf{X} \leq x)$. Let $p : x_i \mapsto p(x_i)$ be a function defined by*

$$p(x_i) = \begin{cases} F_{\mathbf{X}}(x_1) & \text{for} & i = 1 \\ F_{\mathbf{X}}(x_i) - F_{\mathbf{X}}(x_{i-1}) & \text{for} & 1 < i. \end{cases}$$

Then p is a probability distribution on the range of the random variable $\mathbf{X}$, and moreover

$$F_{\mathbf{X}}(x) = \begin{cases} 0 & \text{if} \quad x < x_1 \\ \sum_{x_i \le x} p(x_i) & \text{if} \quad x \ge x_1. \end{cases}$$

Conversely, let $p : \omega \mapsto p(\omega)$ be a probability distribution for some probability space $(\Omega, \mathcal{F}, \mathbb{P})$, with Ω finite or countable. Let $\mathbf{X}$ be any random variable. For any real number x, we define

$$S(x; \mathbf{X}) = \{\omega \in \Omega \,|\, \mathbf{X}(\omega) \le x\}.$$

We also define

$$F_{\mathbf{X}}(x) = \sum_{\omega \in S(x;\mathbf{X})} p(\omega).$$

Then $F_{\mathbf{X}}$ is the distribution function of the r.v. $\mathbf{X}$.

In the uncountable case, a similar relationship exists (in some situations) between a distribution function and a continuous analogue of the probability distribution called the 'density function.' As we shall see in the next chapter, the summation in Eq. (13.3) is then replaced by an integral.

Problems

1. Nero is taking a multiple-choice test, but since he has not studied, he is randomly guessing one of the four answers on each question, with each guess independent. Let $\mathbf{U}$ be the random variable which takes the value 1 if he gets the first question correct and 0 otherwise, and let $\mathbf{V}$ be the random variable giving the number of answers he gets correct in the first three questions. (a) Display the graphs of the distribution functions $F_{\mathbf{U}}$ and $F_{\mathbf{V}}$. (b) Is $F_{\mathbf{U}} + F_{\mathbf{V}}$ a distribution function? How about $\frac{1}{2}(F_{\mathbf{U}} + F_{\mathbf{V}})$?

2. Archie takes the same multiple-choice test, also guessing (independently) on each question. Suppose he answers questions until he gets one correct. Graph the distribution function of the random variable that represents the question number on which his correct guess occurs.

3. Let $\mathbf{Y}$ be the random variable which gives the sum when two fair dice are thrown (cf. Example 12.1). Construct (or define) the distribution function $F_{\mathbf{Y}}$.

4. Let $\mathbf{Z}_n$ be the random variable which gives the number of heads observed in n tosses of a fair coin. Define the distribution function $F_{\mathbf{Z}_n}$ and compare the expressions of $F_{\mathbf{Z}_n}$ for different values of n.

5. Consider **Y** and $\mathbf{Z}_n$ as in Problems 3 and 4 above. (a) Is $F_\mathbf{Y}^2$ a distribution function? How about $F_{\mathbf{Z}_n}^2$? (b) Can you state and prove a general result in this connection? (Is there something special about the 'squaring' function?)

6. Let **T** and **V** be discrete, jointly distributed random variables. Find a relationship between $F_{\mathbf{V+T}}$ and $F_\mathbf{T}$.

7. Let **Z** be the random variable of Example 13.3. (a) $F_\mathbf{Z}(\mathbf{Z})$ is a random variable. Why? (b) Construct (or define) the distribution function $F_\mathbf{Z}(\mathbf{Z})$. (c) What do you notice about $F_\mathbf{Z}(\mathbf{Z})$? Is there a general result lurking about here?

8. Prove Theorem 13.4.

9. Give an interpretation of the following function $f : \mathbb{R} \rightarrow \mathbb{R}$ and its integral $F(x) = \int_{-\infty}^x f(y)dy$, with

$$f(x) = \begin{cases} 0 & \text{for } x < 0 \text{ or } x > 1 \\ 2 & \text{for } 0 \leq x < .1 \\ .5 & \text{for } x = .1 \\ \frac{1}{3} & \text{for } .1 \leq x \leq 1. \end{cases}$$

Chapter 14

Continuous Random Variables

In the case of a discrete sample space Ω, the distribution function $F_{\mathbf{X}}$ of any random variable $\mathbf{X}$ can be defined from the probability distribution p on Ω. Specifically, we obtained

$$F_{\mathbf{X}}(x) = \sum_{\omega \in \mathcal{S}(x, \mathbf{X})} p(\omega) \qquad (14.1)$$

with $\mathcal{S}(x, \mathbf{X}) = \{\omega \in \Omega \mid \mathbf{X}(\omega) \leq x\}$ (cf. Theorem 13.4). A similar situation arises with uncountable sample spaces when the distribution function of $\mathbf{X}$ is differentiable. The probability distribution is then replaced by a 'density' function, and the summation in (14.1) by an integral. The purpose of this chapter is to give the relevant definitions and to discuss the important examples of the 'uniform' or 'rectangular' distribution, the 'exponential' distribution, and the 'normal' or 'Gaussian' distribution. These are but three of a fairly vast collection of distributions playing a prominent role in probability and statistics.

Densities

14.1. Definition. Let $\mathbf{X}$ be a random variable with a distribution function $F_{\mathbf{X}}$. Suppose that $F_{\mathbf{X}}(x)$ is differentiable everywhere except possibly a finite number of points. The random variable $\mathbf{X}$ is then said to be *continuous*, and its (*probability*) *density function* (or more

115

simply, *density*) $f_{\mathbf{X}}$ is defined by

$$f_{\mathbf{X}}(x) = \begin{cases} F'_{\mathbf{X}}(x) = \frac{dF_{\mathbf{X}}}{dx}(x) & \text{if } F'_{\mathbf{X}} \text{ is defined at } x \\ 0 & \text{otherwise.} \end{cases} \tag{14.2}$$

We have then, for any real number x

$$F_{\mathbf{X}}(x) = \int_{-\infty}^{x} f_{\mathbf{X}}(y)\, dy \tag{14.3}$$

a result analoguous to (14.1). If $F_{\mathbf{X}}$ is continuous, then necessarily, for all $x \in \mathbb{R}$,

$$\mathbb{P}(\mathbf{X} = x) = \mathbb{P}(\mathbf{X} \leq x) - \lim_{y \to x-} \mathbb{P}(\mathbf{X} \leq y) = F_{\mathbf{X}}(x) - \lim_{y \to x-} F_{\mathbf{X}}(y)$$

$$= F_{\mathbf{X}}(x) - F_{\mathbf{X}}(x)$$

$$= 0.$$

We have thus, for any two real numbers $x_1 < x_2$,

$$\mathbb{P}(x_1 \leq \mathbf{X} \leq x_2) = \mathbb{P}(x_1 < \mathbf{X} \leq x_2)$$
$$= \mathbb{P}(x_1 \leq \mathbf{X} < x_2) = \mathbb{P}(x_1 < \mathbf{X} < x_2)$$
$$= F_{\mathbf{X}}(x_2) - F_{\mathbf{X}}(x_1) = \int_{x_1}^{x_2} f_{\mathbf{X}}(y)\, dy. \tag{14.4}$$

(compare with Eq. (13.2)). Whereas the values of a probability distribution p represent probabilities,[1] the values of a density function $f_{\mathbf{X}}$ do not. In fact, its definition does not prevent $f_{\mathbf{X}}(x)$ from being greater than 1 for some point(s). (For an example, see Fig. 14.3 (a) in this chapter). Rather, as indicated by Eq. (14.4), the definite integral of the density function represents a probability.

We shall illustrate the concepts of this definition with three important examples.

The Uniform (or Rectangular) Distribution

14.2. Empirical Situation. You are waiting for a train, which is known to always arrive between time a and time b. Moreover, the likelihood that the train arrives within the interval $]x, x + \delta[$ does not depend on x (for x and δ satisfying $a < x < x + \delta < b$).

---

[1] For example, $p(x)$ represents the probability of the event $\{x\}$.

14.3. Definition. Let $\mathbf{X}$ be a continuous random variable with density function $f_{\mathbf{X}}$. We say that $\mathbf{X}$ is a *uniform* (or *rectangular*) random variable if there are two real numbers $a < b$ such that

$$f_{\mathbf{X}}(x) = \begin{cases} \frac{1}{b-a} & \text{if } a < x < b \\ 0 & \text{otherwise.} \end{cases} \tag{14.5}$$

In such a case, we also say, equivalently, that $\mathbf{X}$ has a *uniform* (or *rectangular*) distribution with parameters a and b. The distribution function $F_{\mathbf{X}}$ of $\mathbf{X}$ has then the form

$$F_{\mathbf{X}}(x) = \begin{cases} 0 & \text{if } x \le a \\ \frac{x-a}{b-a} & \text{if } a < x < b \\ 1 & \text{if } b \le x. \end{cases} \tag{14.6}$$

A graph of the distribution function and of the density function of a uniform random variable with parameters 2 and 4.5 is pictured in Fig. 14.1.

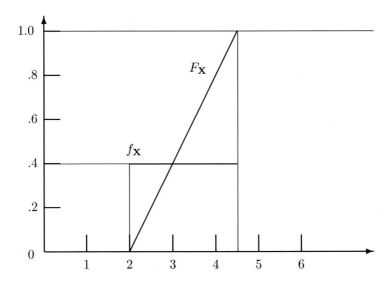

Figure 14.1: Graphs of the density function $f_{\mathbf{X}}$ and the distribution function $F_{\mathbf{X}}$ of a uniform r.v. with parameters 2 and 4.5 (see Definition 14.3).

The next theorem describes a remarkable property of uniform random variables.

14.4. Theorem. *If* **X** *is a continuous random variable with a strictly increasing distribution function* F, *then the random variable* $F(\mathbf{X})$ *is uniform with parameters 0 and 1.*

PROOF. It is easily seen that $F(\mathbf{X})$ is a random variable. Since F is a distribution function, its range is a subset of the closed interval $[0, 1]$. Consequently, we have $\mathbb{P}(F(\mathbf{X}) \leq t) = 0$ for any $t < 0$ and $\mathbb{P}(F(\mathbf{X}) \leq t) = 1$ for any $t > 1$. For $0 \leq t \leq 1$, we have

$$\mathbb{P}(F(\mathbf{X}) \leq t) = \mathbb{P}\left(\mathbf{X} \leq F^{-1}(t)\right) = F\left(F^{-1}(t)\right) = t. \qquad (14.7)$$

We conclude that the function $t \mapsto \mathbb{P}(F(\mathbf{X}) \leq t)$, which is the distribution function of $F(\mathbf{X})$, is indeed uniform with parameters 0 and 1. □

The Exponential Distribution

14.5. Empirical Situation. You are waiting for a train, which is known to always arrive after or at time 0. The waiting phenomenon has the frustrating 'memory-less' property that, if you already have waited t units of time, the likelihood that the train will arrive within δ units of time, that is, with the interval $]t, t+\delta[$ (with $\delta > 0$) is constant, independent of t. It turns out that this property characterizes the class of distributions defined in this section (see 14.7).

14.6. Definition. Let **T** be a continuous random variable with density function $f_\mathbf{T}$. We say that **T** is an *exponential* random variable if there is a real number $\lambda > 0$ such that

$$f_\mathbf{T}(t) = \begin{cases} \lambda e^{-\lambda t} & \text{if } t \geq 0 \\ 0 & \text{otherwise.} \end{cases} \qquad (14.8)$$

In such a case, we also say, equivalently, that **T** has an *exponential distribution* with parameter $\lambda > 0$. It is easy to show by integration that the distribution function $F_\mathbf{T}$ of **T** has then the form

$$F_\mathbf{T}(t) = \begin{cases} 1 - e^{-\lambda t} & \text{if } t \geq 0 \\ 0 & \text{otherwise.} \end{cases} \qquad (14.9)$$

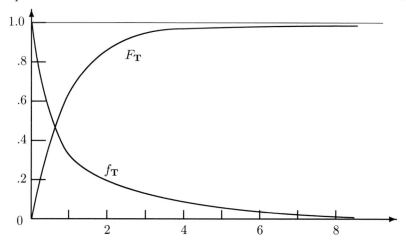

Figure 14.2: Graphs of the density function f_T and the distribution function F_T of an exponential r.v. with parameter $\lambda = 1$ (see Definition 14.6).

A graph of the density function and the distribution function of an exponential random variable with parameter $\lambda = 1$ is displayed in Fig. 14.2.

The memory-less property of the phenomenon described by an exponential random variable is precisely defined by Eq. (14.10) in the next theorem.

14.7. Theorem. *Let $\mathbf{T}$ be a random variable satisfying the following two conditions:*

(i) $\mathbb{P}(\mathbf{T} \leq 0) = 0$;

(ii) *for all $t > 0$ and $\delta > 0$, we have $\mathbb{P}(\mathbf{T} > t) > 0$ and moreover*

$$\mathbb{P}(\mathbf{T} > t + \delta \mid \mathbf{T} > t) = \mathbb{P}(\mathbf{T} > \delta). \qquad (14.10)$$

Then, $\mathbf{T}$ is exponential, that is, we necessarily have, for some constant $\lambda > 0$ and all $t > 0$

$$\mathbb{P}(\mathbf{T} \leq t) = 1 - e^{-\lambda t}. \qquad (14.11)$$

Our proof of Theorem 14.7 is based on the following well-known result from the field of functional equations.

14.8. Lemma. *Let g be a real valued, monotonic function defined on the positive reals, and satisfying $g(x+y) = g(x)g(y)$. Then g has necessarily the form $g(x) = e^{Kx}$ for some real constant K.*

For a proof of this lemma see, for example, Aczél (1966).[2]

PROOF OF THEOREM 14.7. By definition of conditional probability, we have

$$\mathbb{P}(\mathbf{T} > t + \delta \,|\, \mathbf{T} > t) = \frac{\mathbb{P}(\mathbf{T} > t + \delta, \mathbf{T} > t)}{\mathbb{P}(\mathbf{T} > t)}$$

$$= \frac{\mathbb{P}(\mathbf{T} > t + \delta)}{\mathbb{P}(\mathbf{T} > t)} \qquad \text{(because }]t + \delta, \infty[\subseteq \,]t, \infty[\,)$$

$$= \mathbb{P}(\mathbf{T} > \delta) \qquad \text{(by Eq. (14.10))}.$$

Multiplying by $\mathbb{P}(\mathbf{T} > t)$ in the last equation, and writing $g(t) = \mathbb{P}(\mathbf{T} > t)$, we obtain $g(t + \delta) = g(t)g(\delta)$ for all $t, \delta > 0$. By the definition, the function g is monotonic: $t < t'$ implies $g(t) \geq g(t')$. Applying Lemma 14.8, we obtain $g(t) = e^{Kt}$, or equivalently, with $\lambda = -K$

$$\mathbb{P}(\mathbf{T} \leq t) = 1 - g(t) = 1 - e^{-\lambda t}$$

with $\lambda > 0$ because $\lim_{t \to \infty} \mathbb{P}(\mathbf{T} \leq t) = 1$. □

The Normal (or Gaussian) Distribution

It is by all means the most important distribution in probability and statistics. There are several reasons why this is so. Among them, perhaps the most weighty is the so-called 'Central Limit Theorem' which roughly states that if we have a large number of independent and identically distributed random variables, then their sum is approximately distributed normally. (This and related topics are discussed in Chapter 17.) Perhaps as a result of this property, many phenomena can be described by a normal distribution.

14.9. Definition. A random variable $\mathbf{X}$ is said to be *normal* or *Gaussian* with parameters μ and $\sigma > 0$ if its density $f_{\mathbf{X}}$ statisfies the equation

$$f_{\mathbf{X}}(x) = \frac{1}{\sigma \sqrt{2\pi}} e^{-\frac{1}{2}\left(\frac{x-\mu}{\sigma}\right)^2}. \qquad (14.12)$$

---

[2] Aczél's book is the classic exposition of the field of functional equations.

The distribution function

$$F_{\mathbf{X}}(x) = \frac{1}{\sigma\sqrt{2\pi}} \int_{-\infty}^{x} e^{-\frac{1}{2}\left(\frac{y-\mu}{\sigma}\right)^2} dy$$

does not have an analytic form.[3] A normal random variable $\mathbf{Z}$ with parameters $\mu = 0$ and $\sigma = 1$ is referred to as a *standard normal* random variable. In this case, one often denotes the density function and the distribution function by the greek letters ϕ and Φ, respectively, thus,

$$\mathbb{P}(\mathbf{Z} \leq z) = \Phi(z) = \int_{-\infty}^{z} \phi(y)\, dy = \frac{1}{\sqrt{2\pi}} \int_{-\infty}^{z} e^{-\frac{y^2}{2}}\, dy.$$

Three examples of normal density functions are displayed in Fig. 14.3.

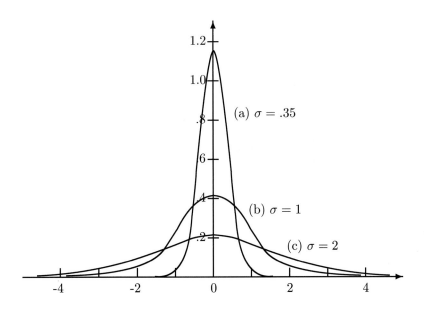

Figure 14.3: Three examples of normal density functions: (a) $\mu = 0$, $\sigma = .35$; **(b)** $\mu = 0$, $\sigma = 1$; **(c)** $\mu = 0$, $\sigma = 2$.

A convenient property of the normal family of distributions is expressed by the following theorem.

---

[3]Many approximation formulas are available (see, for example, Johnson et al., 1994). More-over, due to the importance of the normal distribution in practical applications, most computers have a dedicated routine for evaluating such an approximation.

14.10. Theorem. *If* $\mathbf{X}$ *is a normal random variable, then* $a\mathbf{X}+b$ *is also normal for any real numbers* $a > 0$ *and* b. *In particular, suppose that* $\mathbf{X}$ *has parameters* μ *and* $\sigma > 0$. *Then* $\frac{\mathbf{X}-\mu}{\sigma}$ *is a standard normal random variable, that is, it has parameters* 0 *and* 1. *In such a case, we have for any real numbers* $x_1 > x_2$,

$$\mathbb{P}(x_1 \le \mathbf{X} \le x_2) = F_{\mathbf{X}}(x_2) - F_{\mathbf{X}}(x_1)$$
$$= \Phi(\frac{x_2 - \mu}{\sigma}) - \Phi(\frac{x_1 - \mu}{\sigma}), \qquad (14.13)$$

where $F_{\mathbf{X}}$ *and* Φ *are the distribution functions of* $\mathbf{X}$ *and of a standard normal random variable, respectively.*

14.11. Remark. In the proof, we use the following fact from calculus concerning the change of variable in integration: If f and g are continuous functions on the real interval $[x_1, x_2]$ with g strictly increasing and g^{-1} is the inverse function of g, then

$$\int_{x_1}^{x_2} f(x)\,dx = \int_{g(x_1)}^{g(x_2)} f[g^{-1}(y)]\,dg^{-1}(y). \qquad (14.14)$$

Note that the integral in the right-hand side of (14.14) is a *Riemann-Stieltjes integral* of the form $\int_a^b h(x)\,d\alpha(x)$,[4] which is defined for any real valued functions h and α on a closed interval $[a, b]$, with h continuous and α increasing (see any advanced calculus or analysis text, for example, Rudin, 1964).

PROOF OF THEOREM 14.10. In the equations below, we change the variable as in Eq. (14.14) with $f = f_{\mathbf{X}}$, the density function of $\mathbf{X}$, and $g(x) = \frac{x-\mu}{\sigma}$. We have successively

$$\mathbb{P}(x_1 \le \mathbf{X} \le x_2) = \frac{1}{\sigma\sqrt{2\pi}} \int_{x_1}^{x_2} e^{-\frac{1}{2}\left(\frac{y-\mu}{\sigma}\right)^2}\,dy$$

$$= \frac{1}{\sigma\sqrt{2\pi}} \int_{\frac{x_1-\mu}{\sigma}}^{\frac{x_2-\mu}{\sigma}} e^{-\frac{y^2}{2}}\,d(\sigma y + \mu)$$

$$= \frac{1}{\sqrt{2\pi}} \int_{\frac{x_1-\mu}{\sigma}}^{\frac{x_2-\mu}{\sigma}} e^{-\frac{y^2}{2}}\,dy$$

$$= \Phi(\frac{x_2 - \mu}{\sigma}) - \Phi(\frac{x_1 - \mu}{\sigma}).$$

$\square$

---

[4]Often abbreviated as $\int_a^b h\,d\alpha$.

Problems

1. Archie always leaves the old brownstone on 35th Street at exactly 5:50 a.m. to go to Penn Station and take the train to Long Island. Let the random variable $\mathbf{T}$ represent the time (in minutes) that Archie takes to reach the train's track. The train always leaves exactly at 6 a.m. What is the probability that Archie is on the train on any particular day, if $\mathbf{T}$ is

   (a) uniform with parameters 0 and 15?

   (b) exponential with parameter $\lambda = 0.1$?

   (c) normal with parameters $\mu = 9$ and $\sigma = 1$?

2. Suppose that $\mathbf{T}$ as above is exponential with parameter $\lambda = 0.1$. What is the probability that Archie catches the train, given that he still has not arrived at the train's track at 5:54 a.m.?

3. Suppose that the time it takes for a car to be serviced at a service station is exponentially distributed with parameter λ. Suppose also that there are only two garages at the station, so that at most two cars may be serviced at one time. If Cars 1, 2, and 3 arrive at the station at the same time, with Cars 1 and 2 entering the garages and Car 3 waiting for one of these cars to be serviced before entering a garage, what is the probability that Car 3 will be serviced before one of the other two cars? What is the probability that Car 1 will be serviced last?

4. Let $\mathbf{X_1}$ and $\mathbf{X_2}$ be independent random variables, each uniformly distributed with parameters 0 and 1. The maximum of $\mathbf{X_1}$ and $\mathbf{X_2}$ is a random variable (see 12.16). Find its distribution and density functions.

5. If $\mathbf{Y_1}$ and $\mathbf{Y_2}$ are independent, exponential random variables with parameters λ_1 and λ_2, respectively, show that their minimum is exponential with parameter $\lambda_1 + \lambda_2$.

6. Let $\mathbf{X}$ be a continuous random variable whose range is a subset of the positive reals and whose density and distribution functions are respectively given by $f_\mathbf{X}$ and $F_\mathbf{X}$. Compute the function $\frac{f_\mathbf{X}}{1 - F_\mathbf{X}}$ when $\mathbf{X}$ is exponential with parameter λ. Using your answer, try to give an interpretation of the function $\frac{f_\mathbf{X}}{1 - F_\mathbf{X}}$. (Hint: for the last part, the function $\frac{f_\mathbf{X}}{1 - F_\mathbf{X}}$ is called the *hazard function* or the *(conditional) failure rate function* of the random variable $\mathbf{X}$.)

7. Compute $\mathbb{P}(1 < e^\mathbf{X} < 2)$ in two cases: (a) $\mathbf{X}$ is distributed normally with parameters $\mu = 1$ and $\sigma = 2$; (b) $\mathbf{X}$ is distributed exponentially with parameter $\lambda = 1$. In Case (b), describe completely the distribution of $e^\mathbf{X}$.

Chapter 15

Expectation and Moments

Considerable information on the distribution of a random variable can be obtained from its 'moments'. One important moment is the 'expectation' (or mean) of the random variable, which locates the 'center of gravity' of the distribution on the real line. (We will be more specific later on in this chapter.) Consider, for example, a random variable $\mathbf{Y}$ with a binomial distribution p and parameters α and n. The expectation of $\mathbf{Y}$ is equal to

$$n\alpha = \sum_{k=0}^{n} k \binom{n}{k} \alpha^k (1-\alpha)^{n-k} \qquad (15.1)$$

(See Problem 1). In this case, and this applies in fact to any discrete random variable, the expectation is the sum of all the values of the random variable, each value being weighted by its probability.

The situation is similar for continuous random variables. For a normal random variable $\mathbf{X}$ with parameters μ and $\sigma > 0$, for example, it can be shown by integration that

$$\mu = \frac{1}{\sigma\sqrt{2\pi}} \int_{-\infty}^{\infty} x\, e^{-\frac{1}{2}\left(\frac{x-\mu}{\sigma}\right)^2} dx, \qquad (15.2)$$

and μ is called the expectation of $\mathbf{X}$. The expectation of a r.v. is its 'first moment'. Higher moments capture other aspects of the distribution. For instance, the value of the parameter σ indicates the 'spread' of the distribution. (The three examples of Fig. 14.3 illustrate this point.) It turns out that σ^2 is the 'central moment of order 2' of $\mathbf{X}$ called the 'variance' of $\mathbf{X}$. We shall see that the third moment (which involves the power '3') measures the asymm the distribution of a random variable. This chapter and the next one contain the most important definitions and facts regarding the concepts of moments and (in particular) expectation of a random variable.

Expectation of a Function of a Random Variable

We will define the 'expectation of a function G of a random variable $\mathbf{X}$.' This must be understood as follows: A random variable $\mathbf{X}$ is a mapping of the sample space into the reals. Thus, any point ω of the sample space is mapped to some real number $\mathbf{X}(\omega)$. Suppose that we apply some real valued function G to the result, to yield $G(\mathbf{X}(\omega))$. We have thus the diagram

$$\omega \xmapsto{\ \mathbf{X}\ } \mathbf{X}(\omega) \xmapsto{\ G\ } G[\mathbf{X}(\omega)].$$

If G is continuous, then $G(\mathbf{X}(.)) = G \circ \mathbf{X}$ is also a random variable (with $\circ$ denoting the composition of functions). By the 'expectation of the function G of the random variable $\mathbf{X}$,' we simply mean the expectation of the random variable $G(\mathbf{X})$. This concept is defined below.

15.1. Definition. Let $\mathbf{X}$ be a random variable in a probability space $(\Omega, \mathcal{F}, \mathbb{P})$, let $F_{\mathbf{X}}$ be the distribution function of $\mathbf{X}$, and let G be any real valued, continuous function on $\mathbb{R}$. (Thus, $G(\mathbf{X})$ is a random variable.) The *expectation* of $G(\mathbf{X})$ is defined by

$$E\left(G(\mathbf{X})\right) = \begin{cases} \sum_{\omega \in \Omega} G\left(\mathbf{X}(\omega)\right) p_{\mathbf{X}}(\omega) & \text{if } \mathbf{X} \text{ is discrete with} \\ & \text{probability distribution } p_{\mathbf{X}}; \\ \int_{-\infty}^{\infty} G(x) f_{\mathbf{X}}(x)\, dx & \text{if } f_{\mathbf{X}} = F'_{\mathbf{X}}. \end{cases}$$

Using the Riemann-Stieltjes Integral (cf. Remark 14.11), we can gather these two cases into the more general formula

$$E\left(G(\mathbf{X})\right) = \int_{-\infty}^{\infty} G(x)\, dF_{\mathbf{X}}(x). \tag{15.3}$$

shall sometimes use this type of notation because it conveniently discrete and the continuous case. (If any perplexing udent should translate it into one involving a discrete case, with a probability distribution and a summation sign.) Note that the expectation in Eq. (15.3) does not necessarily exist, that is, there may not be a real number corresponding to $E\left(G(\mathbf{X})\right)$. In an infinite sample space, this expectation is defined as a series (in the discrete case) or as an integral (in the continuous case). The point is that this series or this integral does not necessarily converge. An example is given in 15.11. Two examples of (15.3) are given below.

15.2. Example. Let $\mathbf{Y}$ be a Bernoulli-distributed r.v. with parameter $\alpha = .25$, and let $G(x) = (x - 3)^2$ for any real number x. Then $E\left(G(\mathbf{Y})\right) = .75(0 - 3)^2 + .25(1 - 3)^2 = 1.675$.

15.3. Example. Let $\mathbf{T}$ be an exponentially distributed r.v. with parameter $\lambda \neq 1$, and let $G(\mathbf{T}) = e^{\mathbf{T}}$. Then

$$E\left(G(\mathbf{T})\right) = \int_0^\infty e^t \, \lambda \, e^{-\lambda t} dt = \lambda \int_0^\infty e^{-(\lambda - 1)t} dt = \frac{\lambda}{\lambda - 1}.$$

Many examples of this concept will be found in this chapter. Note that the expectation is a 'linear operator' in the sense of the next theorem.

15.4. Theorem. *Let $\mathbf{X}$ and $\mathbf{Y}$ be two jointly distributed random variables. Provided that the expectations converge, we have for any two real numbers a and b,*

$$E(a\mathbf{X} + b\mathbf{Y}) = aE(\mathbf{X}) + bE(\mathbf{Y}). \tag{15.4}$$

PROOF. First, notice that by 12.14, $a\mathbf{X} + b\mathbf{Y}$ is a random variable. We give a proof for a discrete sample space Ω with probability distribution p. (The proof in the continuous case is similar; cf. Problem 14 in Chapter 19.) Notice that for any point ω in Ω, we have $(a\mathbf{X} + b\mathbf{Y})(\omega) = a\mathbf{X}(\omega) + b\mathbf{Y}(\omega)$ by definition of the addition of two functions. Equation (15.4) follows from the string of equalities

$$E(a\mathbf{X} + b\mathbf{Y}) = \sum_{\omega \in \Omega} (a\mathbf{X} + b\mathbf{Y})(\omega)p(\omega)$$

$$= \sum_{\omega \in \Omega} [a\mathbf{X}(\omega) + b\mathbf{Y}(\omega)]p(\omega)$$

$$= a \sum_{\omega \in \Omega} \mathbf{X}(\omega)p(\omega) + b \sum_{\omega \in \Omega} \mathbf{Y}(\omega)p(\omega)$$

$$= aE(\mathbf{X}) + bE(\mathbf{Y}). \qquad \square$$

15.5. Remark. By convention, we also give a meaning to expressions such as $E(a)$, where a is a real constant. Implicitly, $E(a)$ means $E(a \cdot \mathbf{I})$ where $\mathbf{I}$ denotes a random variable[1] having 'all its mass concentrated at 1,' that is, with $\mathbb{P}(\mathbf{I} = 1) = 1$. This implies

$$E(a) = E(a \cdot \mathbf{I}) = aE(\mathbf{I}) = a.$$

The Moments of a Random Variable

The expectation (or mean) of a random variable $\mathbf{X}$ is the special case of Def. 15.1 where the function G in (15.3) is the identity function. This is the 'moment of order one' of $\mathbf{X}$. A general definition is given below.

15.6. Definition. Let $\mathbf{X}$ be a random variable with distribution function $F_{\mathbf{X}}$, and let n be any natural number. The expectation of the r.v. $\mathbf{X}^n$, that is,

$$E(\mathbf{X}^n) = \int_{-\infty}^{\infty} x^n \, dF_{\mathbf{X}}(x) \tag{15.5}$$

is called the *moment of order n* of $\mathbf{X}$. The moment of order 1 is called the *expectation* or the *mean* of $\mathbf{X}$. For the rest of this definition, we write $\mu = E(\mathbf{X})$. The *central moment of order n* of $\mathbf{X}$ is defined by

$$E[(\mathbf{X} - \mu)^n] = \int_{-\infty}^{\infty} (x - \mu)^n \, dF_{\mathbf{X}}(x). \tag{15.6}$$

The central moment of order 2 is called the *variance* of $\mathbf{X}$, and we write

$$Var(\mathbf{X}) = E[(\mathbf{X} - \mu)^2]. \tag{15.7}$$

The variance is a standard index of the spread of the distribution. The following expression of the variance in terms of the first two moments is important for both theoretical and practical reasons:

$$Var(\mathbf{X}) = E(\mathbf{X}^2) - E(\mathbf{X})^2. \tag{15.8}$$

---

[1]For the readers of the starred sections, the random variable $\mathbf{I}$ is simply I_Ω, that is, the indicator function of the sample space, which is a random variable; see 2.9 and Problem 10 in Chapter 12.

By Theorem 15.4, we have indeed

$$Var(\mathbf{X}) = E[(\mathbf{X}-\mu)^2] = E(\mathbf{X}^2+\mu^2-2\mu\mathbf{X}) = E(\mathbf{X}^2)+\mu^2-2\mu E(\mathbf{X}).$$

Frequently, what is of interest is $\sqrt{Var(\mathbf{X})}$, the square root of the variance, because it has the same unit as $\mathbf{X}$. This index is referred to as the *standard deviation* of the r.v. $\mathbf{X}$.

15.7. Example. We compute the first two moments of an exponential random variable $\mathbf{T}$ with parameter λ. By integration, we obtain

$$E(\mathbf{T}) = \int_0^\infty t\,\lambda e^{-\lambda t}dt = \frac{1}{\lambda}, \qquad E(\mathbf{T}^2) = \int_0^\infty t^2\,\lambda e^{-\lambda t}dt = \frac{2}{\lambda^2}$$

which, using (15.8), gives

$$Var(\mathbf{T}) = E(\mathbf{T}^2) - E(\mathbf{T})^2 = \frac{2}{\lambda^2} - \left(\frac{1}{\lambda}\right)^2 = \frac{1}{\lambda^2}. \qquad (15.9)$$

In practice, one rarely derives the moments by integration. The important concept of the 'moment generating function' is introduced in the next chapter. We shall see that, if the moment generating function (M.G.F.) of a random variable is available, then all the moments can be obtained by computing the successive derivatives of this function.

Table 15.1 contains, for useful reference, the means and the variances of all the basic distributions encountered in Chapters 8 and 14, as well as other information on these r.v.'s such as their probability distributions or density functions, and the expressions of their moment generation functions, the meaning of which will be made clear in Chapter 18.

15.8. Remark. Notice that, for any random variable $\mathbf{X}$ and any two real constants a and b, we have

$$Var(a\mathbf{X} + b) = a^2 Var(\mathbf{X}). \qquad (15.10)$$

Indeed,

$$\begin{aligned}
Var(a\mathbf{X} + b) &= E\left((a\mathbf{X} + b)^2\right) - (E(a\mathbf{X} + b))^2 \\
&= E(a^2\mathbf{X}^2 + b^2 + 2ab\mathbf{X}) - (aE(\mathbf{X}) + b)^2 \\
&= a^2 E(\mathbf{X}^2) + b^2 + 2ab E(\mathbf{X}) - a^2 E(\mathbf{X})^2 - b^2 - 2ab E(\mathbf{X}) \\
&= a^2 Var(\mathbf{X}).
\end{aligned}$$

Table 15.1: Names and main features of some common distributions. The functional form of the distribution is given in the third column of the Table, either by its probability distribution (rows 2-5), or by its density (rows 6-8). The M.G.F. is discussed in Chapter 18.

Name	Parameters	Distribution	Mean	Variance	M.G.F.
Bernoulli	$\alpha \in [0,1]$	$p(1)=1-p(0)=\alpha$	α	$\alpha(1-\alpha)$	$\alpha e^\theta + 1 - \alpha$
Binomial	$n \in \mathbb{N}, \alpha \in [0,1]$	$p(k)=\binom{n}{k}\alpha^k(1-\alpha)^{n-k}$ $k=0,1,\dots,n$	$n\alpha$	$n\alpha(1-\alpha)$	$(\alpha e^\theta + 1 - \alpha)^n$
Geometric	$\alpha \in]0,1[$	$p(k)=\alpha(1-\alpha)^{k-1}$ $k=0,1,\dots,n,\dots$	$\frac{1}{\alpha}$	$\frac{1-\alpha}{\alpha^2}$	$\frac{\alpha e^\theta}{1-(1-\alpha)e^\theta}$
Poisson	$\lambda > 0$	$p(k)=e^{-\lambda}\frac{\lambda^k}{k!}$ $k=0,1,2,\dots$	λ	λ	$e^{\lambda(e^\theta-1)}$
Uniform	$-\infty < a < b < \infty$	$f(x)=\frac{1}{b-a}, a<x<b$ $=0,$ elsewhere	$\frac{a+b}{2}$	$\frac{(b-a)^2}{12}$	$\frac{e^{\theta b}-e^{\theta a}}{\theta(b-a)}$
Exponential	$\lambda > 0$	$f(t)=\lambda e^{-\lambda t}, t>0$ $=0,$ elsewhere	$\frac{1}{\lambda}$	$\frac{1}{\lambda^2}$	$\frac{\lambda}{\lambda-\theta}$
Normal	$\sigma > 0, \mu \in \mathbb{R}$	$f(x)=\frac{1}{\sigma\sqrt{2\pi}}e^{-\frac{1}{2}\left(\frac{x-\mu}{\sigma}\right)^2}$	μ	σ^2	$e^{\theta\mu+(1/2)\theta^2\sigma^2}$

15.9. Remark. With $\mu = E(\mathbf{X})$, the third central moment of the random variable $\mathbf{X}$ is $E\left((\mathbf{X}-\mu)^3\right)$. This quantity is a measure of the asymmetry of the distribution of a random variable $\mathbf{X}$. A normalization is often carried out, in order to control for the effect of the unit of $\mathbf{X}$. (A good index of asymmetry for a random variable $\mathbf{X}$ measuring the weights of individuals should not depend upon whether the weights are expressed in pounds or in kilograms.)

A standard index for a random variable $\mathbf{X}$ with expectation μ and Variance $\sigma^2 > 0$ is

$$As(\mathbf{X}) = \frac{\left(E\left((\mathbf{X}-\mu)^3\right)\right)^{\frac{1}{3}}}{\sigma}.$$

It is easy to verify that $As(a\mathbf{X}+b) = As(\mathbf{X})$ if $a > 0$.

15.10. Remark. The mean of a random variable does not necessarily exist. As pointed out in 15.1, in an infinite sample space, an expectation is defined as the limit of a series or as an integral, and this series or this integral does not necessarily converge. This remark applies to the mean and to other moments. A simple example is given below.

15.11. Example. Take the set $\mathbb{N}$ of natural numbers as the sample space, and let p be the geometric distribution with parameter $\alpha = \frac{1}{2}$. Thus $p(n) = \left(\frac{1}{2}\right)^n$ and $\sum_{n=1}^{\infty} p(n) = 1$. We define on $\mathbb{N}$ the function $\mathbf{W}(n) = 2^n$. Then $\mathbf{W}$ is a discrete random variable with probability distribution

$$
\mathbb{P}(\mathbf{W} = x) = \begin{cases} \frac{1}{2^n} & \text{if } x = 2^n \text{ with } n \in \mathbb{N}, \\ 0 & \text{otherwise.} \end{cases}
$$

The expectation of $\mathbf{W}$ does not converge: We have

$$
E(\mathbf{W}) = \sum_{n=1}^{\infty} 2^n \cdot \left(\frac{1}{2}\right)^n = 1 + 1 + \ldots + 1 + \ldots
$$

15.12. Remark. In the context of game theory, the above example is cast as the ST. PETERBURG'S PARADOX. Imagine a game between two players, with the following rules. A fair coin is tossed repeatedly until the first head occurs. If the first head occurs on the nth trial, Player A gives 2^n dollars to Player B. However, to be allowed to play this game, Player B must initially give a certain sum to Player A. We can safely venture that few people would give more than, say, \$100 to play this game. However, as shown in the example above, the expected gain of Player B is infinite.

Other Indices of Central Location

Besides the mean, there are two other standard indices of the location of the 'center' of the distribution of a random variable.

15.13. Definition. A number m_e is called a *median* of the distribution of a random variable $\mathbf{X}$ if

$$
\mathbb{P}(\mathbf{X} \leq m_e) = \frac{1}{2}.
$$

Clearly, a median always exists but may not be unique. On the other hand, if the distribution function of the r.v. is strictly increasing, then there is a unique median.

A *mode* of the distribution of a discrete r.v. $\mathbf{X}$ is a number m_o at which the probability distribution has a maximum. In other words,

m_o is a mode if for all real numbers x

$$\mathbb{P}(\mathbf{X} = m_o) \geq \mathbb{P}(\mathbf{X} = x).$$

The definition in the continuous case is similar: m_o is a mode if it is a maximum of the density function of $\mathbf{X}$. As in the case of the median, a mode always exists but may not be unique.

In some situations, there is a unique mode and a unique median which coincide with the mean. An example is any normal random variable. On the other hand, any number between 0 and 1 is technically a mode for a uniform random variable with parameters 0 and 1. In such a case, the concept of mode is useless.

Problems

1. Give two different proofs that the expectation of a binomially distributed random variable with parameters α and n is equal to $n\alpha$ (cf. Eq. (15.1)). (Do not use the moment generating function.)

2. Consider a situation in which n letters are randomly placed into n envelopes (with each envelope getting one letter). Suppose that each letter has a certain envelope in which it belongs. What is the expected number of letters that end up in the correct envelopes?

3. The baseball World Series is played between two teams, with the series ending when one of the teams wins four games. Find the expectation and the variance of the number of games played, assuming that each team is equally likely to win a game and that the games' outcomes are independent.

4. An airport bus has four possible stop points. Suppose that two passengers take the bus. Suppose also that a passenger is equally likely to get off at any of the four stops, with the two passengers acting independently. The bus stops only if someone wants to get off. What are the expectation and the variance of the number of stops that the bus makes?

5. Is it the case in general that the expected value of the product of two random variables is equal to the product of their expected values? Justify your answer.

6. Give an example of a random variable $\mathbf{X}$ having a mean, but no variance (that is, $Var(\mathbf{X})$ is not a real number).

7. Compute the mean and the variance of the geometric, Poisson, and uniform distributions.

8. Check that for $a, b \in \mathbb{R}$, we have $|a|As(a\mathbf{X} + b) = aAs(\mathbf{X})$ (cf. Remark 15.9). Can two distributions on the same sample space have the same mean and the same variance, but different values of the index As? If your answer is positive, give a simple example.

9. Compute the median of an exponential distribution having $\lambda = 3$ as the parameter.

10. Could there be a discrete random variable with countably many modes?

11. Show that $E[(\mathbf{X} - c)^2]$, as a function of the real variable c, is minimum for $c = E(\mathbf{X})$. (Suppose that the expectation converges.)

12. Can you think of any other index of the asymmetry of a distribution (cf. Remark 15.9)?

Chapter 16

Covariance and Correlation

Two nonindependent random variables may be related in various forms and degrees. We begin by considering a standard index of such covariation.

Covariance

16.1. Definition. The *covariance* of two jointly distributed random variables $\mathbf{X}$ and $\mathbf{Y}$ is defined by

$$Cov(\mathbf{X}, \mathbf{Y}) = E(\mathbf{XY}) - E(\mathbf{X})E(\mathbf{Y}). \qquad (16.1)$$

Notice that by Eq. (15.8), we have $Cov(\mathbf{X}, \mathbf{X}) = Var(\mathbf{X})$. Also, using Theorem 15.4, we get[1]

$$E(\mathbf{XY}) - E(\mathbf{X})E(\mathbf{Y}) = E\left((\mathbf{X} - E(\mathbf{X}))\left(\mathbf{Y} - E(\mathbf{Y})\right)\right),$$

and so,

$$Cov(\mathbf{X}, \mathbf{Y}) = E\left((\mathbf{X} - E(\mathbf{X}))\left(\mathbf{Y} - E(\mathbf{Y})\right)\right). \qquad (16.2)$$

In some situations, the covariance is a good index of the degree of dependence between two random variables. The three matrices in Table 16.1 give some intuition on the results of the computation of a covariance in some exemplary cases.

In Case (a), we have $E(\mathbf{XY}) = 0$ and $E(\mathbf{X}) = E(\mathbf{Y}) = \frac{1}{2}$, yielding $Cov(\mathbf{X}, \mathbf{Y}) = E(\mathbf{XY}) - E(\mathbf{X})E(\mathbf{Y}) = -\frac{1}{4}$, a negative number indicating that the high values of $\mathbf{X}$ tend to coincide with the low values of $\mathbf{Y}$, and vice versa.

---

[1]The student should check the algebra. Note that, to lighten the language, in the statement of results concerning expectations, we implicitly assume that the relevant expectations exist.

Table 16.1: Three examples for the computation of the covariance .

		Y	
		0	1
X	1	$\frac{1}{2}$	0
	0	0	$\frac{1}{2}$

(a)

		W	
		0	1
V	0	0	$\frac{1}{2}$
	-1	$\frac{1}{2}$	0

(b)

		Z		
		-1	0	1
T	1	0	0	$\frac{1}{4}$
	0	$\frac{1}{4}$	$\frac{1}{4}$	0
	-1	0	0	$\frac{1}{4}$

(c)

A similar computation for Case (b) gives us

$$Cov(\mathbf{V}, \mathbf{W}) = \frac{1}{4}$$

a positive number indicating that the high values of $\mathbf{V}$ and $\mathbf{W}$ tend to occur together, and that the low values of these two r.v.'s also tend to occur together.

However, note that in Case (c), we have $E(\mathbf{TZ}) = E(\mathbf{T}) = 0$, which gives

$$Cov(\mathbf{T}, \mathbf{Z}) = 0$$

even though there is an obvious lack of independence between $\mathbf{T}$ and $\mathbf{Z}$. For example, we have

$$\mathbb{P}(\mathbf{T} = 1, \mathbf{Z} = 1) = \frac{1}{4} > \mathbb{P}(\mathbf{T} = 1)\mathbb{P}(\mathbf{Z} = 1) = \frac{1}{4} \cdot \frac{1}{2} = \frac{1}{8}.$$

In general, we have the following result:

16.2. Theorem. *If two jointly distributed r.v.'s* $\mathbf{X}$ *and* $\mathbf{Y}$ *are independent, then* $Cov(\mathbf{X}, \mathbf{Y}) = 0$. *In general, the converse implication does not hold. However, if both* $\mathbf{X}$ *and* $\mathbf{Y}$ *are normally distributed, then they are independent if and only if* $Cov(\mathbf{X}, \mathbf{Y}) = 0$.

PARTIAL PROOF. We only prove the first statement in the discrete case (see Theorem 20.5). Suppose that $\mathbf{X}$ and $\mathbf{Y}$ are discrete, independent random variables, and let $S_\mathbf{X}$ and $S_\mathbf{Y}$ be the ranges of the functions $\mathbf{X}$ and $\mathbf{Y}$, respectively. Thus, $S_\mathbf{X}$ and $S_\mathbf{Y}$ are at most countable. Successively, using Eq. (12.13),

$$
\begin{aligned}
Cov(\mathbf{X}, \mathbf{Y}) &= \sum_{x \in S_\mathbf{X}} \sum_{y \in S_\mathbf{Y}} xy \, \mathbb{P}(\mathbf{X} = x, \mathbf{Y} = y) - E(\mathbf{X})E(\mathbf{Y}) \\
&= \sum_{x \in S_\mathbf{X}} \sum_{y \in S_\mathbf{Y}} xy \, \mathbb{P}(\mathbf{X} = x)\mathbb{P}(\mathbf{Y} = y) - E(\mathbf{X})E(\mathbf{Y}) \\
&= E(\mathbf{X})E(\mathbf{Y}) - E(\mathbf{X})E(\mathbf{Y}) = 0.
\end{aligned}
$$

Case (c) of Table 16.1 proves that the converse does not hold. ☐

Proofs of the first statement in the continuous case, and of the second statement: *for normally distributed random variables,*

$$\mathbf{X} \text{ and } \mathbf{Y} \text{ are independent} \iff Cov(\mathbf{X}, \mathbf{Y}) = 0$$

can be based on a two-dimensional analogue of the density function of a normal random variable, that is, the 'joint density' of $\mathbf{X}$ and $\mathbf{Y}$, a concept discussed later in this monograph (see Theorem 20.6).

The second statement in the next theorem is an important consequence of Theorem 16.2.

16.3. Theorem. *For any jointly distributed random variables* $\mathbf{X}$ *and* $\mathbf{Y}$, *we have*

$$Var(\mathbf{X} + \mathbf{Y}) = Var(\mathbf{X}) + Var(\mathbf{Y}) + 2Cov(\mathbf{X}, \mathbf{Y}). \qquad (16.3)$$

Moreover, if $\mathbf{X}$ *and* $\mathbf{Y}$ *are independent, then*

$$Var(\mathbf{X} + \mathbf{Y}) = Var(\mathbf{X}) + Var(\mathbf{Y}). \qquad (16.4)$$

PROOF. We have, by definition of the variance

$$
\begin{aligned}
Var(\mathbf{X} + \mathbf{Y}) &= E[(\mathbf{X} + \mathbf{Y})^2] - E(\mathbf{X} + \mathbf{Y})^2 \\
&= E(\mathbf{X}^2) + E(\mathbf{Y}^2) + 2E(\mathbf{XY}) - E(\mathbf{X})^2 \\
&\qquad - E(\mathbf{Y})^2 - 2E(\mathbf{X})E(\mathbf{Y}) \\
&= Var(\mathbf{X}) + Var(\mathbf{Y}) + 2Cov(\mathbf{X}, \mathbf{Y}),
\end{aligned}
$$

after rearranging, by definition of the covariance. Equation (16.4) results from (16.3) and Theorem 16.2. ☐

16.4. Remark. The measure of the dependence existing between two random variables $\mathbf{X}$ and $\mathbf{Y}$ provided by the covariance depends of course of the particular scale used to measure the phenomena which are numerically represented by these random variables. Suppose, for example, that someone is interested in the relationship between the length of time spent in the United States, and the body weight, for a particular subpopulation of recent immigrants. The particular value of the covariance between the two corresponding random variables will depend on the units used for the duration of the stay (years, months, or weeks) and the body weight (pounds or kilograms). Indeed, it is easily shown that, for any real constants a, b, c, and d, we have

$$Cov(a\mathbf{X} + b, c\mathbf{Y} + d) = acCov(\mathbf{X}, \mathbf{Y}). \qquad (16.5)$$

To control for such spurious scale factors, a normalization of the covariance is carried out, which we introduce in the following section.

The Correlation Coefficient

16.5. Definition. The *(Pearson) correlation coefficient* between two random variables is the function

$$\rho_{\mathbf{X},\mathbf{Y}} = \frac{Cov(\mathbf{X}, \mathbf{Y})}{\sqrt{Var(\mathbf{X})Var(\mathbf{Y})}}, \qquad (16.6)$$

which is defined for any two random variables $\mathbf{X}$ and $\mathbf{Y}$ provided that the relevant expectations exist and the variances in the denominator do not vanish. When $\rho_{\mathbf{X},\mathbf{Y}} = 0$, we may say that $\mathbf{X}$ and $\mathbf{Y}$ are *uncorrelated*. As could be expected, we have the following result:

16.6. Theorem. *If $\mathbf{X}$ and $\mathbf{Y}$ are two random variables, and a, b, c, and d are real numbers with $ac \neq 0$, we have*

$$\rho_{a\mathbf{X}+b,c\mathbf{Y}+d} = \frac{ac}{|ac|}\rho_{\mathbf{X},\mathbf{Y}}.$$

This follows immediately from Eq. (16.5) and the fact that when a r.v. is multiplied by some constant k, its variance is multiplied by k^2 (cf. Remark 15.8). Thus, the value of the correlation coefficient does not change if a and c are both positive or both negative, and changes sign in the other cases. In particular, we have $\rho_{\mathbf{X},a\mathbf{X}+b} = 1$, for any two real numbers b and $a > 0$. Our next theorem requires a preparatory result.

16.7. Lemma. *For any two r.v.'s* $\mathbf{V}$ *and* $\mathbf{W}$, *we always have*

$$-\sqrt{E(\mathbf{V}^2)E(\mathbf{W}^2)} \le E(\mathbf{VW}) \le \sqrt{E(\mathbf{V}^2)E(\mathbf{W}^2)}. \tag{16.7}$$

PROOF. *Case a.* Suppose that one of $E(\mathbf{V}^2)$ or $E(\mathbf{W}^2)$ is equal to 0, say $E(\mathbf{V}^2) = 0$. This implies $E(\mathbf{V})^2 = 0$ since $Var(\mathbf{V}) = E(\mathbf{V}^2) - E(\mathbf{V})^2 \ge 0$. But then $\mathbb{P}(\mathbf{V} = 0) = 1$, and because

$$\{\omega \in \Omega \,|\, \mathbf{V}(\omega) = 0\} \subseteq \{\omega \in \Omega \,|\, \mathbf{V}(\omega)\mathbf{W}(\omega) = 0\},$$

we conclude that $\mathbb{P}(\mathbf{VW} = 0) = 1$. Thus, $E(\mathbf{VW}) = 0$, and (16.7) follows.

Case b. Suppose that neither $E(\mathbf{V}^2)$ nor $E(\mathbf{W}^2)$ vanish. We have then

$$0 \le E\left[\left(\frac{\mathbf{V}}{\sqrt{E(\mathbf{V}^2)}} + \frac{\mathbf{W}}{\sqrt{E(\mathbf{W}^2)}}\right)^2\right]$$

$$= \frac{E(\mathbf{V}^2)}{E(\mathbf{V}^2)} + \frac{E(\mathbf{W}^2)}{E(\mathbf{W}^2)} + 2\frac{E(\mathbf{VW})}{\sqrt{E(\mathbf{V}^2)E(\mathbf{W}^2)}}$$

$$= 2 + 2\frac{E(\mathbf{VW})}{\sqrt{E(\mathbf{V}^2)E(\mathbf{W}^2)}},$$

which implies, after some algebraic manipulations,

$$-\sqrt{E(\mathbf{V}^2)E(\mathbf{W}^2)} \le E(\mathbf{VW}), \tag{16.8}$$

the left inequality in (16.7). Similarly, we get

$$0 \le E\left[\left(\frac{\mathbf{V}}{\sqrt{E(\mathbf{V}^2)}} - \frac{\mathbf{W}}{\sqrt{E(\mathbf{W}^2)}}\right)^2\right] = 2 - 2\frac{E(\mathbf{VW})}{\sqrt{E(\mathbf{V}^2)E(\mathbf{W}^2)}}$$

leading to

$$E(\mathbf{VW}) \le \sqrt{E(\mathbf{V}^2)E(\mathbf{W}^2)}. \tag{16.9}$$

The result in Case b follows from (16.8) and (16.9). $\qquad\square$

16.8. Theorem. *Let* $\mathbf{X}$ *and* $\mathbf{Y}$ *be two random variables having variances* $\sigma_{\mathbf{X}}^2 > 0$ *and* $\sigma_{\mathbf{Y}}^2 > 0$, *respectively. We have then*

$$-1 \le \rho_{\mathbf{X},\mathbf{Y}} \le 1.$$

This follows immediately from Lemma 16.7 by setting

$$\mathbf{V} = \mathbf{X} - E(\mathbf{X}), \qquad \mathbf{W} = \mathbf{Y} - E(\mathbf{Y}),$$

and dividing Eq. (16.7) by $\sqrt{E[(\mathbf{X} - E(\mathbf{X}))^2]E[(\mathbf{Y} - E(\mathbf{Y}))^2]}$.

Problems

1. Concoct another example of a pair of nonindependent random variables having a correlation coefficient equal to 0.

2. Verify Eq. (16.5).

3. Let $\mathbf{V}$, $\mathbf{W}$, and $\mathbf{T}$ be random variables having a common variance σ^2. Compute $\rho_{\mathbf{V}+\mathbf{W},\mathbf{T}+\mathbf{W}}$.

4. Suppose that three balls, labeled '1', '2', and '3,' are placed in an urn, and that a sample of size two with replacement is drawn. Let $\mathbf{U}$ represent the number on the first ball and $\mathbf{V}$ the sum of the numbers on the two balls. Compute $E(\mathbf{U})$, $E(\mathbf{V})$, $Var(\mathbf{U})$, $Var(\mathbf{V})$, and $Cov(\mathbf{U},\mathbf{V})$.

5. Consider the empirical situation described in 8.6, that is, the one involving the multinomial distribution. Let $\mathbf{M}_i$ be the random variable representing the number of balls of type i that are sampled. Find $E(\mathbf{M}_i)$, $Var(\mathbf{M}_i)$, and $Cov(\mathbf{M}_i,\mathbf{M}_j)$.

6. Show that, for random variables $\mathbf{X}_1, \mathbf{X}_2, \dots, \mathbf{X}_n$,

   (a) $Cov(\sum_{i=1}^n \mathbf{X}_i, \sum_{j=1}^m \mathbf{X}_j) = \sum_{i=1}^n \sum_{j=1}^m Cov(\mathbf{X}_i, \mathbf{X}_j)$;
   (b) $Var(\sum_{i=1}^n \mathbf{X}_i) = \sum_{i=1}^n Var(\mathbf{X}_i) + 2\sum_{i=1}^n \sum_{j<i} Cov(\mathbf{X}_i, \mathbf{X}_j)$.

Chapter 17

The Law of Large Numbers

Suppose that the distribution of some random variable is not known exactly, and only a couple of moments, say the mean and the variance, are available. Despite such poor information, is it nevertheless possible to have some rough idea of the probabilities of observing values lying some distance away from the mean. This question has been a concern of statisticians since the 18th century. The first section of this chapter is devoted to a result found independently by I.J. Bienaymé (1796-1878) in France and P.L. Chebyshev (Čebyšev) (1821-1894) in Russia and referred to as 'Bienaymé-Chebyshev Inequality' or, more often, as 'Chebyshev Inequality.'

Chebyshev Inequality

We begin by a more general result (see also Problem 10).

17.1. Theorem. *Let* $\mathbf{X}$ *be a random variable, and let* G *be a continuous, nonnegative real valued function on* $\mathbb{R}$. *Suppose that the expectation* $E\left(G(\mathbf{X})\right)$ *exists and is positive. Then, for any real number* $k > 0$, *we have*

$$\mathbb{P}\big(G(\mathbf{X}) \geq kE(G(\mathbf{X}))\big) \leq \frac{1}{k}. \qquad (17.1)$$

PROOF. Let $S = \{x \in \mathbb{R} \mid G(x) \geq kE[G(\mathbf{X})]\}$ and let $\bar{S} = \mathbb{R} \setminus S$. We also write F for the (unknown) distribution function of $\mathbf{X}$. Using the facts that $G \geq 0$ and that $x \in S$ implies $G(x) \geq kE\left(G(\mathbf{X})\right)$, we get successively

$$E\left(G(\mathbf{X})\right) = \int_{-\infty}^{\infty} G(x)\, dF(x)$$

$$= \int_{S} G(x)\, dF(x) + \int_{\bar{S}} G(x)\, dF(x)$$

$$\geq \int_{S} G(x)\, dF(x)$$

$$\geq kE\left(G(\mathbf{X})\right) \int_{S} dF(x)$$

$$= kE\left(G(\mathbf{X})\right) \mathbb{P}\big(G(\mathbf{X}) \geq kE\left(G(\mathbf{X})\right)\big),$$

by definition of S and F. Thus,

$$E\left(G(\mathbf{X})\right) \geq kE\left(G(\mathbf{X})\right) \mathbb{P}\big(G(\mathbf{X}) \geq kE\left(G(\mathbf{X})\right)\big),$$

and (17.1) follows after dividing both sides by $kE\left(G(\mathbf{X})\right)$. □

Chebyshev Inequality is a special case of Eq. (17.1) in which $G(\mathbf{X}) = (\mathbf{X} - \mu)^2$ and $k = h^2$. We have:

17.2. Theorem. (CHEBYSHEV.)   *Let $\mathbf{X}$ be a random variable, with $E(\mathbf{X}) = \mu$ and $Var(\mathbf{X}) = \sigma^2 > 0$. Then, for any real number $h > 0$, we have*

$$\mathbb{P}(|\mathbf{X} - \mu| \geq h\sigma) \leq \frac{1}{h^2}. \tag{17.2}$$

Notice that since h is an arbitrarily chosen positive real number, we can take $h = \frac{\epsilon}{\sigma}$, which yields the equivalent form

$$\mathbb{P}(|\mathbf{X} - \mu| \geq \epsilon) \leq \frac{\sigma^2}{\epsilon^2}. \tag{17.3}$$

PROOF. Taking $G(\mathbf{X}) = (\mathbf{X} - \mu)^2$ and $k = h^2$ in (17.1), we get

$$\mathbb{P}\big((\mathbf{X} - \mu)^2 \geq h^2 E\left((\mathbf{X} - \mu)^2\right)\big) \leq \frac{1}{h^2}. \tag{17.4}$$

But $E[(\mathbf{X} - \mu)^2] = \sigma^2$ and moreover

$$\{\omega \in \Omega \mid (\mathbf{X}(\omega) - \mu)^2 \geq h^2\sigma^2\} = \{\omega \in \Omega \mid |\mathbf{X}(\omega) - \mu| \geq h\sigma\}.$$

Equation (17.4) is thus equivalent to Eq. (17.2). □

The bounds provided by Chebyshev Inequality are rather crude, as compared to the result of an exact computation when the distribution is known. We give two examples.

17.3. Example. Take the case of a standard normal r.v. $\mathbf{Z}$; thus, $E(\mathbf{Z}) = 0$ and $Var(\mathbf{Z}) = 1$. Using a table for the standard normal[1] or a statistical package, we can compute that $\mathbb{P}(|\mathbf{Z}| \geq 2) \approx .0445$, whereas Chebyshev Inequality gives $\mathbb{P}(|\mathbf{Z}| \geq 2) \leq .25$. The discrepancy is huge.

17.4. Example. Consider a uniform random variable $\mathbf{Y}$ with parameters b and a. We get thus

$$E(\mathbf{Y}) = \frac{b+a}{2} = 0, \qquad Var(\mathbf{Y}) = \frac{(b-a)^2}{12} = \sigma^2.$$

Solving the last equation for the value of $b - a$ in terms of σ gives $b - a = 2\sigma\sqrt{3}$. This leads to

$$\mathbb{P}(|\mathbf{Y}| \geq \frac{3}{2}\sigma) = 1 - \frac{6\sigma}{2(b-a)} = 1 - \frac{3}{2\sqrt{3}} \approx .134.$$

On the other hand, using the Chebyshev Inequality, we get

$$\mathbb{P}(|\mathbf{Y}| \geq \frac{3}{2}\sigma) \leq \frac{4}{9} \approx .444.$$

Here again, the Chebyshev bound is far from the exact value. As we shall shortly see, the inequality is nevertheless important for theoretical purposes.

The Law of Large Numbers

To numerous generations of scientists and gamblers, it has seemed reasonable to estimate the probability of a phenomenon by the relative frequency of its occurrence. For concreteness, consider the probability α of drawing a black ball in a single draw from an urn containing only black and white balls. If α is unknown, a natural estimate of α is the ratio n_b/n, where n_b is the number of black balls obtained in a sample of size n, with replacement. In fact, we have seen in 11.2 that n_b/n was a maximum likelihood estimate

---

[1]Which is available in any elementary statistics text.

of α. Common sense and experience agree to suggest that such an estimate would be increasingly accurate as the total number n of draws increases. The first general result confirming this conjecture is attributed to Jakob Bernoulli (1654-1705) and appeared posthumously in his book *Ars Conjectandi* (Bernoulli, 1713).

Until fairly recently (1900-1930), the difficulty in handling this type of problem was due to the lack of appropriate tools. The proper setting is as a sequence of independent Bernoulli random variables $\mathbf{X}_1, \ldots, \mathbf{X}_n, \ldots$, with a common parameter α and for $n = 1, 2, \ldots$

$$\mathbf{X}_n = \begin{cases} 1 & \text{if a black ball is drawn on trial } n \\ 0 & \text{otherwise.} \end{cases}$$

In this context, the relative frequency of occurrence of a black ball over n draws is the random variable

$$\overline{\mathbf{X}}_n = \frac{\mathbf{X}_1 + \mathbf{X}_2 + \ldots + \mathbf{X}_n}{n}. \tag{17.5}$$

An intuitive interpretation of the conjecture is that $\overline{\mathbf{X}}_n$ converges in some sense to α. Note that, because $\overline{\mathbf{X}}_n$ is a random variable for any positive integer n whereas α is real number, it is not readily clear what kind of convergence is involved. A more general question arises if we drop the assumption that the $\mathbf{X}_n$ are Bernoulli r.v.'s, or even have the same distribution. The random variable $\overline{\mathbf{X}}_n$ is then simply the mean of a set of values successively sampled from the independent random variables $\mathbf{X}_n$ having the same mean, and we conjecture that the sample mean $\overline{\mathbf{X}}_n$ converges in some sense to the common mean μ. The result below, known as the 'Law of Large Numbers', specifies the type of convergence. We suppose here, however, that the random variables have the same variance.

17.5. Theorem. (LAW OF LARGE NUMBERS.) *Suppose that* $\mathbf{X}_1$, $\ldots, \mathbf{X}_n, \ldots$ *are independent r.v.'s with the same expectation* μ *and the same variance* σ^2, *and let* $\overline{\mathbf{X}}_n$ *be defined by Eq. (17.5). Then*

$$\lim_{n \to \infty} \mathbb{P}(|\overline{\mathbf{X}}_n - \mu| \geq \epsilon) = 0. \tag{17.6}$$

If the r.v.'s $\mathbf{X}_n$ are assumed to be Bernoulli-distributed, then the above Theorem becomes 'Bernoulli's Law of Large Numbers.' In the proof, we use the following fact, which is an immediate consequence of 15.8 and 16.3: If $\mathbf{X}_1, \mathbf{X}_2, \ldots, \mathbf{X}_n$ are independent random variables with the same variance σ^2, then

$$Var\left(\frac{\mathbf{X}_1 + \mathbf{X}_2 + \ldots + \mathbf{X}_n}{n}\right) = \frac{\sigma^2}{n}. \qquad (17.7)$$

PROOF. Using Chebyshev Inequality in the form Eq. (17.3), we obtain

$$\mathbb{P}(|\overline{\mathbf{X}}_n - \mu| \geq \epsilon) \leq \frac{\sigma^2}{n\epsilon^2}.$$

As $\lim_{n \to \infty} \frac{\sigma^2}{n\epsilon^2} = 0$, the theorem follows. $\qquad \square$

17.6. Remark. This theorem says nothing about the speed of convergence, which depends on the distributions of the random variables $\mathbf{X}_n$. What it says is that, no matter what these distributions are, the distribution of $\overline{\mathbf{X}}_n$ will eventually collapse around μ. If, however, we can assume that all the random variables $\mathbf{X}_n$ have the same distribution, then much more than the Law of Large Numbers can be asserted, namely, for even moderately large n (say, ≥ 30), the distribution of $\overline{\mathbf{X}}_n$ is approximately normal, with mean μ. The next section is devoted to a fundamental result in this regard.

The Central Limit Theorem

17.7. Definition. Let $\mathbf{X}$ be a random variable, and let $\mathbf{X}_1, \ldots, \mathbf{X}_n$ be n independent random variables having all the same distribution as $\mathbf{X}$. Then, the set $\{\mathbf{X}_1, \ldots, \mathbf{X}_n\}$ is called a *random sample (of size n)* of the r.v. $\mathbf{X}$, and the random variable

$$\overline{\mathbf{X}}_n = \frac{\mathbf{X}_1 + \ldots + \mathbf{X}_n}{n}$$

is called the *sample mean.*

Suppose that $\mathbf{X}$ has a finite mean and finite variance, specifically,

$$E(\mathbf{X}) = \mu, \qquad\qquad Var(\mathbf{X}) = \sigma^2;$$

then, it follows easily that

$$E(\overline{\mathbf{X}}_n) = \mu, \qquad\qquad Var(\overline{\mathbf{X}}_n) = \frac{\sigma^2}{n}.$$

Thus, as n gets large, the distribution of $\overline{\mathbf{X}}_n$ gets concentrated around μ. This is the gist of the Law of Large Numbers. The Central Limit Theorem is concerned with the distribution of $\overline{\mathbf{X}}_n$. It states, in essence, that the distribution of $\overline{\mathbf{X}}_n$ is approximately normal for large n. Notice that, in the statement of this result given in the theorem below, a normalization of $\overline{\mathbf{X}}_n$ is carried out so as to obtain a random variable with a mean equal to 0 and a variance equal to 1. We have indeed (cf. Problem 4)

$$E\left(\frac{\overline{\mathbf{X}}_n - \mu}{\sigma/\sqrt{n}}\right) = 0, \qquad Var\left(\frac{\overline{\mathbf{X}}_n - \mu}{\sigma/\sqrt{n}}\right) = 1. \qquad (17.8)$$

17.8. Theorem. (THE CENTRAL LIMIT THEOREM.)   *For any $n \in \mathbb{N}$, let $\{\mathbf{X}_1, \dots, \mathbf{X}_n\}$ be a random sample from a random variable $\mathbf{X}$ having a finite mean μ and a finite variance σ^2. Then, for any real numbers $a \leq b$, we have*

$$\lim_{n\to\infty} \mathbb{P}\left(a \leq \frac{\overline{\mathbf{X}}_n - \mu}{\sigma/\sqrt{n}} \leq b\right) = \Phi(b) - \Phi(a), \qquad (17.9)$$

where Φ denotes the distribution function of a standard normal random variable.

Note that this limit does not depend upon the distribution of $\mathbf{X}$. The proof of this Theorem is beyond the scope of this book. This result plays a fundamental role in probability theory and in statistics, and justifies the prominent place of the normal distribution in these fields.

It must be pointed out that the convergence is fast, even when $\mathbf{X}$ has a very skewed distribution. In practical applications, $n \geq 30$ is regarded as sufficient to ensure the approximate normality of $\overline{\mathbf{X}}_n$ (or of the normalized form used in Eq. (17.9)).

17.9. Example. Suppose that $\mathbf{X}$ has an exponential distribution with parameter $\lambda = 1$. Figure 17.1 displays the densities of the sum of random variables $\mathbf{X}_1 + \dots + \mathbf{X}_n = n\overline{\mathbf{X}}_n$ in four random samples of $\mathbf{X}$ of sizes 1, 2, 3 and 10. With $n = 10$, to the eye, the sum already appears almost symmetric.

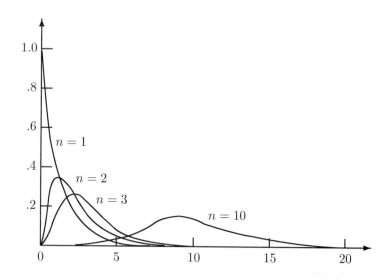

Figure 17.1: Plots of the density functions of four sums $n\bar{X}_n$, with $n = 1, 2, 3, 10$, of an exponential random variable X with parameter $\lambda = 1$.

Problems

1. Apply Theorem 17.1 for **X**, an exponentially distributed random variable with parameter $\lambda = 2$, $G(\mathbf{X}) = e^{\mathbf{X}}$, and $k = 2$. Compare the result with the exact computation (cf. Example 15.3).

2. Apply Bienaymé-Chebyshev Inequality for an exponential random variable with $\lambda = 2$. Compare the result with the exact computation.

3. Does the Law of Large Numbers still hold if the random variables are uncorrelated rather than independent? If your answer is yes, prove the Theorem under this new assumption.

4. Verify the two equations in (17.8).

5. Suppose that the amount of time Archie waits at the post office is a random variable distributed exponentially with mean six minutes. Suppose also that the waiting times are independent from visit to visit and that Archie makes a total of ten trips to the post office.

   (a) Use Theorem 17.1 to obtain a bound on the probability that Archie waits at least an hour and a half combined.

   (b) Use the Central Limit Theorem to estimate the probability that he waits at least an hour and a half combined.

6. Suppose that $\mathbf{X}_1, \mathbf{X}_2, \dots, \mathbf{X}_n$ are independent random variables with common mean μ and variance σ^2. Show that $Cov(\bar{\mathbf{X}}_n, \mathbf{X}_i - \bar{\mathbf{X}}_n) = 0$ for $1 \le i \le n$.

7. What is the maximum probability that a random variable is at least n standard deviations from its mean?

8. Calculate the approximate probability that, in 100 tosses of a fair die, the number '6' appears at most ten times.

9. Let $\mathbf{X}$ be a random variable whose mass is concentrated on the nonnegative reals and whose expectation is $\mu < \infty$. Is it necessarily the case that $\mathbb{P}(\mathbf{X} \ge 1) \le \mu$?

*10. (Cf. Loève, 1963, p. 157.) Let $\mathbf{X}$ be a finite, nonnegative random variable in some probability space with sample space Ω, and let $g : [0, \infty[\to [0, \infty[$ be a nonconstant and nondecreasing function. Prove that, for every $a \ge 0$, with $A = \{x \in \Omega \mid \mathbf{X}(x) \ge a\}$,

$$\mathbb{P}(\mathbf{X} \ge a) \ge \frac{E\left(g(\mathbf{X})\right) - g(a)}{\max g\left(\mathbf{X}(A)\right)}.$$

(Hint: Examine the proof of Theorem 17.1.)

Chapter 18

Moment Generating Functions

The 'moment generating function' of a random variable $\mathbf{X}$ is a powerful tool for the study of its distribution. In particular, all the moments of $\mathbf{X}$ can be obtained from its M.G.F. by taking successive derivatives of it, which is an almost automatic operation. Essentially, the M.G.F. of a random variable uniquely specifies its distribution function[1] and greatly facilitates some derivations.

Definition and Examples

We begin with the general definition.

18.1. Definition. The *moment generating function* or *M.G.F.* of a random variable $\mathbf{X}$ is the function

$$M_{\mathbf{X}}(\theta) = E(e^{\theta \mathbf{X}}), \qquad (18.1)$$

which is assumed to be defined for any real number θ in an open interval containing 0. (The reasons for this specification will become clear in a moment.) The notation of the M.G.F., with θ in the parentheses and $\mathbf{X}$ as an index correctly suggests that, for most purposes, θ is the 'working variable,' while $\mathbf{X}$ is fixed.

---

[1]In some rare cases that we need not worry about here, the M.G.F. is not defined because the expectation in Eq. (18.1) does not converge (see Remark 18.11 however).

18.2. Example. Let **Y** be a Poisson random variable with parameter λ. Thus,

$$\mathbb{P}(\mathbf{Y} = k) = \begin{cases} e^{-\lambda}\dfrac{\lambda^k}{k!} & \text{if } k = 0, 1, \dots \\ 0 & \text{otherwise} \end{cases}$$

(cf. Table 15.1). Let us compute $M_{\mathbf{Y}}(\theta)$. Successively, we have

$$E(e^{\theta \mathbf{Y}}) = \sum_{k=0}^{\infty} e^{\theta k} e^{-\lambda} \frac{\lambda^k}{k!}$$

$$= e^{-\lambda} e^{e^{\theta}\lambda} \sum_{k=0}^{\infty} e^{-(e^{\theta}\lambda)} \frac{(e^{\theta}\lambda)^k}{k!}$$

$$= e^{\lambda(e^{\theta}-1)}$$

because the last series converges to 1. (Indeed, $e^{-(e^{\theta}\lambda)}(e^{\theta}\lambda)^k/k!$ can be regarded as the expression of a probability distribution of a Poisson random variable with parameter $e^{\theta}\lambda$.) We conclude that

$$M_{\mathbf{Y}}(\theta) = e^{\lambda(e^{\theta}-1)}, \qquad (18.2)$$

which is the expression in the last column of Table 15.1 for the Poisson distribution.

18.3. Example. Let **V** be a normal random variable with mean μ and standard deviation σ. Then

$$M_{\mathbf{V}}(\theta) = \frac{1}{\sigma\sqrt{2\pi}} \int_{-\infty}^{\infty} e^{\theta x} e^{-\frac{1}{2}\left(\frac{x-\mu}{\sigma}\right)^2} dx$$

$$= \frac{1}{\sigma\sqrt{2\pi}} \int_{-\infty}^{\infty} e^{\theta(\sigma y + \mu)} e^{-\frac{1}{2}y^2} d(\sigma y + \mu)$$

(applying the transformation $x \mapsto \frac{x-\mu}{\sigma} = y$; cf. Remark 14.11)

$$= \frac{e^{\theta\mu}}{\sqrt{2\pi}} \int_{-\infty}^{\infty} e^{-\frac{1}{2}(y^2 - 2\theta\sigma\mu y)} dy$$

$$= \frac{e^{\theta\mu} e^{\frac{1}{2}\theta^2\sigma^2}}{\sqrt{2\pi}} \int_{-\infty}^{\infty} e^{-\frac{1}{2}(y^2 - 2\theta\sigma\mu y + \theta^2\sigma^2)} dy$$

(completing the square in the exponent)

$$= e^{\theta\mu + \frac{1}{2}\theta^2\sigma^2} \left[\frac{1}{\sqrt{2\pi}} \int_{-\infty}^{\infty} e^{-\frac{1}{2}(y - \theta\sigma)^2} dy \right] = e^{\theta\mu + \frac{1}{2}\theta^2\sigma^2},$$

which yields

$$M_{\mathbf{V}}(\theta) = e^{\theta\mu + \frac{1}{2}\theta^2\sigma^2} \tag{18.3}$$

(cf. the lower right cell of Table 15.1).

Computing the Moments

As suggested by its name, an important use of the M.G.F. is to provide the moments of a random variable by a straightforward operation. The working theorem for this purpose is given below.

18.4. Theorem. *Let $M_{\mathbf{X}}(\theta)$ be the M.G.F. of a random variable $\mathbf{X}$ with distribution function $F_{\mathbf{X}}$, and let $M_{\mathbf{X}}^{(n)}(\theta)$ be the nth derivative of the function $M_{\mathbf{X}}$ with respect to θ. Then*

$$M_{\mathbf{X}}^{(n)}(\theta) = \int_{-\infty}^{\infty} x^n e^{\theta x} dF_{\mathbf{X}}(x), \tag{18.4}$$

and thus

$$E(\mathbf{X}^n) = M_{\mathbf{X}}^{(n)}(\theta)|_{\theta=0} = M_{\mathbf{X}}^{(n)}(0). \tag{18.5}$$

Thus, the nth moment of the random variable $\mathbf{X}$ is obtained by computing the nth derivative of the M.G.F. with respect to the dummy variable θ, and setting θ equal to 0 in the resulting expression.

Clearly, Eq. (18.5) immediately follows from Eq. (18.4). (A student uneasy about the proof given below should rewrite it for the case of discrete distributions.)

PROOF. We establish (18.4) by induction. Computing the first derivative of $M_{\mathbf{X}}(\theta)$, we get

$$\frac{d}{d\theta}E(e^{\theta\mathbf{X}}) = \frac{d}{d\theta}\int_{-\infty}^{\infty} e^{\theta x} dF_{\mathbf{X}}(x)$$

$$= \int_{-\infty}^{\infty} \left(\frac{d}{d\theta}e^{\theta x}\right) dF_{\mathbf{X}}(x)$$

$$= \int_{-\infty}^{\infty} x e^{\theta x} dF_{\mathbf{X}}(x).$$

This proves (18.4) for $n = 1$. Suppose that (18.4) holds for $n = k$, and consider the case $n = k + 1$. We have

$$\frac{d}{d\theta} M_{\mathbf{X}}^{(k)}(\theta) = \frac{d}{d\theta} \int_{-\infty}^{\infty} x^k e^{\theta x} dF_{\mathbf{X}}(x)$$

$$= \int_{-\infty}^{\infty} x^k \left(\frac{d}{d\theta} e^{\theta x} \right) dF_{\mathbf{X}}(x)$$

$$= \int_{-\infty}^{\infty} x^{k+1} e^{\theta x} dF_{\mathbf{X}}(x).$$

We have thus

$$M_{\mathbf{X}}^{(k+1)}(\theta) = \int_{-\infty}^{\infty} x^{k+1} e^{\theta x} dF_{\mathbf{X}}(x).$$

Applying induction, we get the theorem.   □

18.5. Example. Pursuing Example 18.2, we compute the mean and the variance of the Poisson distribution from its M.G.F. Using Eq. (18.2), we get $M_{\mathbf{Y}}^{(1)}(\theta) = e^{\lambda(e^\theta - 1)} \lambda e^\theta$, which yields

$$E(\mathbf{Y}) = M_{\mathbf{Y}}^{(1)}(0) = \lambda.$$

We also compute $E(\mathbf{Y}^2)$ by taking the second derivative of $M_{\mathbf{Y}}(\theta)$:

$$E(\mathbf{Y}^2) = M_{\mathbf{Y}}^{(2)}(\theta)|_{\theta=0}$$

$$= \left[e^{\lambda(e^\theta - 1)} \lambda e^\theta + e^{\lambda(e^\theta - 1)} (\lambda e^\theta)^2 \right] \Big|_{\theta=0}$$

$$= \lambda + \lambda^2.$$

We obtain
$$Var(\mathbf{Y}) = E(\mathbf{Y}^2) - E(\mathbf{Y})^2 = \lambda.$$

Besides generating the moments, the M.G.F. has other important features which are summarized in the three theorems of the next section.

Summing Independent Random Variables

18.6. Theorem. *If* $\mathbf{X}$ *and* $\mathbf{Y}$ *are any two independent random variables, then*

$$M_{\mathbf{X}+\mathbf{Y}}(\theta) = M_{\mathbf{X}}(\theta) M_{\mathbf{Y}}(\theta). \tag{18.6}$$

PROOF. By Theorems 12.14 and 12.15, if $\mathbf{X}$ and $\mathbf{Y}$ are independent r.v.'s, then $e^{\theta\mathbf{X}}$ and $e^{\theta\mathbf{Y}}$ are also independent. These r. v.'s thus have a covariance equal to 0 (see Theorem 16.2). We obtain

$$
\begin{aligned}
M_{\mathbf{X}+\mathbf{Y}}(\theta) &= E[e^{\theta(\mathbf{X}+\mathbf{Y})}] && \text{(by definition of the M.G.F.)} \\
&= E(e^{\theta\mathbf{X}}e^{\theta\mathbf{Y}}) \\
&= E(e^{\theta\mathbf{X}})E(e^{\theta\mathbf{Y}}) && \text{(because } Cov(e^{\theta\mathbf{X}}, e^{\theta\mathbf{Y}}) = 0) \\
&= M_{\mathbf{X}}(\theta)M_{\mathbf{Y}}(\theta) && \text{(by definition of the M.G.F.'s).}
\end{aligned}
$$

□

18.7. Example. We apply Theorem 18.6 to compute the M.G.F. of the sum of two independent Poisson random variables $\mathbf{Y}_1$ and $\mathbf{Y}_2$ with parameters λ_1 and λ_2, respectively. Using Eq. 18.2 (or Table 15.1), we obtain

$$
M_{\mathbf{Y}_1+\mathbf{Y}_2}(\theta) = M_{\mathbf{Y}_1}(\theta)M_{\mathbf{Y}_2}(\theta) = e^{\lambda_1(e^\theta-1)}e^{\lambda_2(e^\theta-1)}
$$

which yields

$$
M_{\mathbf{Y}_1+\mathbf{Y}_2}(\theta) = e^{(\lambda_1+\lambda_2)(e^\theta-1)}. \tag{18.7}
$$

The right hand side of this equation has the form of the M.G.F. of a Poisson random variable with parameter $\lambda_1+\lambda_2$. We are tempted to conclude that the sum of two independent Poisson random variables with parameters λ_1 and λ_2 is a Poisson random variable parameter $\lambda_1+\lambda_2$. Because of the theorem below, this conclusion is warranted.[2]

18.8. Theorem. *There is a unique random variable which has a M.G.F. of a given form. In other words, the M.G.F. of a random variable specifies its distribution.*

The proof of this result, which is omitted here, is based on the existence of an 'inversion' formula, which enables the computation of the distribution function of the random variable from the M.G.F.

As demonstrated by Example 18.7, the combination of Theorems 18.6 and 18.8 allows us to determine the distribution of the sum of independent random variables from the individual M.G.F.'s of the components. (Examples are proposed in Problems 5 and 6.)

For another useful and immediate result, we have:

---

[2]The result is actually easy to obtain directly, that is, without relying on Theorem 18.8. (See Problem 8 in this regard.)

18.9. Theorem. *Let* $\mathbf{X}$ *be a random variable, and let* $a \neq 0$ *and* b *be two real numbers. Then*

$$M_{a\mathbf{X}+b}(\theta) = e^{\theta b} M_{\mathbf{X}}(\theta a).$$

18.10. Example. We immediately apply this result to find out the distribution of $a\mathbf{V} + b$, where $\mathbf{V}$ is normally distributed with mean μ and variance σ^2, and a, b are as in Theorem 18.9. Applying this theorem, we get (from Table 15.1),

$$M_{a\mathbf{V}+b}(\theta) = e^{\theta b} e^{\theta a \mu + \frac{1}{2}\theta^2 a^2 \sigma^2}$$

$$= e^{\theta(a\mu+b) + \frac{1}{2}\theta^2 (a\sigma)^2},$$

which is the M.G.F. of a normal r.v. with mean $a\mu + b$ and variance $(a\sigma)^2$. Using Theorem 18.8, we conclude that $a\mathbf{V} + b$ is normally distributed with the same mean and variance.

18.11. Remark. A drawback of the M.G.F. is that, as mentioned in footnote 1 on p. 149, it is not always defined. For that reason, statisticians have devised a closely related complex valued function $\phi_{\mathbf{X}}$ called the *characteristic function* of a random variable $\mathbf{X}$. It is defined, for any random variable $\mathbf{X}$ with distribution function $F_{\mathbf{X}}$, by the equation

$$\phi_{\mathbf{X}}(\theta) = E(e^{i\theta \mathbf{X}}) = \int_{-\infty}^{\infty} e^{i\theta x} dF_{\mathbf{X}}(x),$$

with $i = \sqrt{-1}$ and $e^{i\theta x} = \cos\theta x + i\sin\theta x$. We shall not study the properties of this function here.

Problems

1. Verify Theorem 18.9.

2. Compute the M.G.F. of the following random variables: (a) Bernoulli with parameter α; (b) binomial with parameters α and n; and (c) exponential with parameter λ.

3. Compute the mean and the variance of each of the random variables of Problem 2 via its M.G.F. (cf. Theorem 18.4).

4. The sum of n independent random variables, all distributed exponentially with the same parameter λ, is called a *gamma* random variable with parameters λ and n. Prove that the sum of two independent gamma random variables, one with parameters λ and n and the other with parameters λ and m, is a gamma random variable with parameters λ and $n + m$.

5. Suppose that $\mathbf{X}$ and $\mathbf{Y}$ are independent normal random variables with means $\mu_{\mathbf{X}}$, $\mu_{\mathbf{Y}}$ and variances $\sigma_{\mathbf{X}}^2$, $\sigma_{\mathbf{Y}}^2$, respectively. (a) Show that $a\mathbf{X}+b\mathbf{Y}$, where a and b are nonzero constants, is a normal random variable with mean $a\mu_{\mathbf{X}} + b\mu_{\mathbf{Y}}$ and variance $a^2\sigma_{\mathbf{X}}^2 + b^2\sigma_{\mathbf{Y}}^2$. (b) If $\sigma_{\mathbf{X}} = \sigma_{\mathbf{Y}}$, show that $a\mathbf{X} + b\mathbf{Y}$ and $a\mathbf{X} - b\mathbf{Y}$ are independent.

6. Suppose that $\mathbf{X}$ and $\mathbf{Y}$ are independent binomial r.v.'s, with $\mathbf{X}$ having parameters $\alpha_{\mathbf{X}}$ and m and $\mathbf{Y}$ having parameters $\alpha_{\mathbf{Y}}$ and n. (a) Show that $\mathbf{X} + \mathbf{Y}$ is binomially distributed if $\alpha_{\mathbf{X}} = \alpha_{\mathbf{Y}}$. (b) Is $\mathbf{X} + \mathbf{Y}$ necessarily binomially distributed if $\alpha_{\mathbf{X}} \neq \alpha_{\mathbf{Y}}$?

7. Compute the M.G.F. of a random variable which is the sum of a normal random variable with mean μ and variance σ^2 and an exponential random variable with parameter λ, assuming the two are independent.

8. Let $\mathbf{Y}_1$ and $\mathbf{Y}_2$ be two Poisson random variables with respective parameters λ_1 and λ_2. Prove directly (that is, without relying on Theorem 18.8) that $\mathbf{Y}_1 + \mathbf{Y}_2$ is a Poisson random variable with parameter $\lambda_1 + \lambda_2$.

9. Let $\mathbf{X}_1, \mathbf{X}_2, \ldots, \mathbf{X}_n, \ldots$ be a sequence of independent Poisson r.v.'s, with $\mathbf{X}_n$ having parameter λ^n, $0 < \lambda < 1$. Prove that, as $n \to \infty$, the sequence $\mathbf{S}_n = \sum_{i=1}^{n} \mathbf{X}_i$ tends to a Poisson r.v. with parameter $\frac{\lambda}{1-\lambda}$.

10. Is the sum of independent uniform random variables a uniform random variable? Why or why not?

Chapter 19

Multivariate Distributions

The concept of 'multivariate distribution' may arise whenever we describe a phenomenon by several numerical attributes. In a medical context, one example of such description might be, for each individual in a specified population, the numerical vector

$$(age, \ cholesterol \ level, \ weight, \ blood \ pressure). \qquad (19.1)$$

Each instance of such a vector can be regarded as an outcome in a sample space that can be taken to be $\mathbb{R}^4$. In a discrete case, when each of the variables in the vector runs in a finite or countable subset of $\mathbb{R}$, the multivariate distribution is simply the probability distribution on the set of all such outcomes in the sense of Def.4.5. One example of a discrete multivariate distribution has been encountered implicitly in the guise of the multinomial distribution (cf. Def. 8.7). In the next section, we first deal with this example and then with the discrete case in general.

Notice that the information regarding each attribute can be extracted by 'projecting' the vector on one of its component dimensions. For example, we can define a function

$$\mathbf{X}_1 : (age, \ cholesterol \ level, \ weight, \ blood \ pressure) \mapsto age$$

singling out the age component of the vector. We can obviously proceed similarly with each component of the vector and define the functions $\mathbf{X}_2$, $\mathbf{X}_3$, and $\mathbf{X}_4$ which, in a typical case, turn out to be random variables. Accordingly, it also makes sense to say that multivariate distributions emerge whenever several random variables are defined on the same sample space of some probability space.

The Multinomial as a Multivariate Distribution

19.1. Example. From 8.7, we recall the definition of the multinomial distribution. Let n and k be two integers, with $n \geq k > 0$, and let Ω be a sample space[1] containing all k-tuples $(m_1, \ldots, m_k)$ of nonnegative integers satisfying $\sum_{i=1}^{k} m_i = n$. The multinomial distribution with parameters $\alpha_1, \ldots, \alpha_k \geq 0$ satisfying $\sum_{i=1}^{k} \alpha_i = 1$ is defined by the equation

$$p(m_1, \ldots, m_k) = \binom{n}{m_1 \quad \cdots \quad m_k} \alpha_1^{m_1} \ldots \alpha_k^{m_k}. \qquad (19.2)$$

To cast this distribution as a multivariate distribution, we define, for each index i, with $1 \leq i \leq k$, a function $\mathbf{X}_i : \Omega \to \mathbb{N}$ by the equation

$$\mathbf{X}_i(m_1, \ldots, m_i, \ldots, m_k) = m_i. \qquad (19.3)$$

It should be clear that the functions $\mathbf{X}_i$ are random variables in the probability space induced by Ω and p (cf. Problem 1). The probability distribution p of (19.2) is the joint multivariate distribution of the random variables $\mathbf{X}_1, \ldots, \mathbf{X}_k$, and we can write

$$\mathbb{P}(\mathbf{X}_1 = m_1, \ldots, \mathbf{X}_k = m_k) = \binom{n}{m_1 \quad \cdots \quad m_k} \alpha_1^{m_1} \ldots \alpha_k^{m_k}.$$

19.2. Theorem. *With the functions $\mathbf{X}_i$ defined on Ω as in (19.3), we have, for $1 \leq i \leq k$ and $0 \leq m \leq n$,*

$$\mathbb{P}(\mathbf{X}_i = m) = \binom{n}{m} \alpha_i^m (1 - \alpha_i)^{n-m}. \qquad (19.4)$$

Accordingly, each random variable $\mathbf{X}_i$ has a binomial distribution with parameters α_i and n (cf. 8.4).

PROOF. To simplify the writing, we suppose that $i = 1$. Assume that $\mathbb{P}(\mathbf{X}_1 = m_1) > 0$. We have then

$$\mathbb{P}(\mathbf{X}_2 = m_2, \ldots, \mathbf{X}_k = m_k \mid \mathbf{X}_1 = m_1) = \frac{\mathbb{P}(\mathbf{X}_1 = m_1, \ldots, \mathbf{X}_k = m_k)}{\mathbb{P}(\mathbf{X}_1 = m_1)}.$$

---

[1]One can of course take instead $\mathbb{R}^n$ as the sample space and modify the definition of the probability distribution p accordingly.

Replacing the conditional probability in the left-hand side and the numerator in the right hand side by their multinomial expressions, we get

$$\binom{n-m_1}{m_2 \ldots m_k} \left(\frac{\alpha_2}{1-\alpha_1}\right)^{m_2} \cdots \left(\frac{\alpha_k}{1-\alpha_1}\right)^{m_k}$$
$$= \frac{1}{\mathbb{P}(\mathbf{X}_1 = m_1)} \binom{n}{m_1 \ldots m_k} \alpha_1^{m_1} \ldots \alpha_k^{m_k}.$$

Canceling out the common factors on both sides and solving for $\mathbb{P}(\mathbf{X}_1 = m_1)$ yields (19.4) with $1 = i$ and $m_1 = m$.

We leave the case $\mathbb{P}(\mathbf{X}_1 = m_1) = 0$ to the student (cf. Problem 2).

□

Multivariate Discrete Distributions

The example of the multinomial distribution discussed in the previous section illustrates a general situation described in Def. 19.4. As a preparation for this material, we generalize Theorem 9.8 (Total Probabilities) to the countable case. We omit the proof of the next theorem (see Problem 4).

19.3. Theorem. *Let I be a finite or countable index set, and let $(H_i)_{i \in I}$ be a family of pairwise incompatible events in a probability space $(\Omega, \mathcal{F}, \mathbb{P})$ such that $\sum_{i \in I} H_i = \Omega$. Then, for all $A \in \mathcal{F}$,*

$$\mathbb{P}(A) = \sum_{i \in I} \mathbb{P}(A \cap H_i). \tag{19.5}$$

Moreover, if $\mathbb{P}(H_i) \neq 0$, for all $i \in I$, then,

$$\mathbb{P}(A) = \sum_{i \in I} \mathbb{P}(A \mid H_i) \mathbb{P}(H_i). \tag{19.6}$$

19.4. Definition. Let $\mathcal{X} = (\mathbf{X}_i)_{1 \leq i \leq n}$ be a finite vector of discrete random variables on the same sample space $\mathbb{R}^n$, with probability measure $\mathbb{P}$ (defined thus on the Borel field of $\mathbb{R}^n$). The *multivariate probability distribution* of $\mathcal{X}$ is the function $p_{\mathcal{X}} : \mathbb{R}^n \to [0, 1]$ defined by

$$p_{\mathcal{X}}(x_1, \ldots, x_n) = \mathbb{P}(\mathbf{X}_1 = x_1, \ldots, \mathbf{X}_n = x_n). \tag{19.7}$$

We may also refer to $p_\mathcal{X}$ as the *joint distribution* of the random variables $\mathbf{X}_1, \ldots, \mathbf{X}_n$. Because all these random variables are discrete, the set

$$\mathcal{S} = \{(x_1, \ldots, x_n) \in \mathbb{R}^n \,|\, p_\mathcal{X}(x_1, \ldots, x_n) > 0\}$$

is at most countable. Accordingly, for each real number x and each index i, $1 \le i \le n$, the set $\mathcal{S}_i(x) = \{(x_1, \ldots, x_n) \in \mathcal{S} \,|\, x_i = x\}$ is also most countable, and we can define

$$p_i(x) = \mathbb{P}(\mathbf{X}_i = x) = \sum_{(x_1,\ldots,x_n)\in\mathcal{S}_i(x)} p_\mathcal{X}(x_1, \ldots, x_n). \qquad (19.8)$$

Note that $p_i(x) = 0$ if $\mathcal{S}_i(x) = \emptyset$ for some real number x and that the set $\mathcal{S}_i = \{x \in \mathbb{R} \,|\, \mathcal{S}_i(x) \ne \emptyset\}$ is also at most countable.

For each i, $1 \le i \le n$, we have $p_i \ge 0$ and, with $\mathbf{x} = (x_1, \ldots, x_n)$,

$$\sum_{x\in\mathcal{S}_i} p_i(x) = \sum_{x\in\mathcal{S}_i}\sum_{\mathbf{x}\in\mathcal{S}_i(x)} p_\mathcal{X}(\mathbf{x}) = \sum_{\mathbf{x}\in\mathcal{S}} p_\mathcal{X}(\mathbf{x}) = 1$$

because $p_\mathcal{X}$ is a probability distribution. The function p_i is thus also a probability distribution, which is referred to as the *marginal distribution* of the random variable $\mathbf{X}_i$, corresponding to the joint probability distribution $p_\mathcal{X}$. Note that when only two random variables are considered, we sometimes write $\mathbf{X} = \mathbf{X}_1$ and $\mathbf{Y} = \mathbf{X}_2$ and modify the notation accordingly ($\mathcal{S}_\mathbf{X} = \mathcal{S}_1$, $p_\mathbf{X} = p_1$, etc.).

We exercise this definition in two examples.

19.5. Example. Let p be the probability distribution having a value zero everywhere on $\mathbb{R}^2$, except for its subset $\{1, 2, 3\} \times \{4, 5, 6\}$, where it is defined by Table 19.1.

Table 19.1: Two finite jointly distributed random variables X and Y.

		X			
		1	2	3	$p_\mathbf{Y}$
	6	.10	.05	.15	.30
Y	5	.12	.13	.20	.45
	4	.15	.05	.05	.25
	$p_\mathbf{X}$	.37	.23	.40	1

In the notation of Def. 19.4, we have $\mathcal{S} = \{1, 2, 3\} \times \{4, 5, 6\}$. We define two random variables $\mathbf{X} : (x, y) \mapsto x$ and $\mathbf{Y} : (x, y) \mapsto y$. The marginal distributions of $\mathbf{X}$ and $\mathbf{Y}$ are specified in the bottom row and the last column of Table 19.1. We have, for example,

$$\mathcal{S}_\mathbf{X}(2) = \{(2,4), (2,5), (2,6)\}, \qquad \mathcal{S}_\mathbf{Y}(5) = \{(1,5), (2,5), (3,5)\},$$
$$p_\mathbf{X}(2) = \mathbb{P}(\mathbf{X} = 2) = .23, \qquad p_\mathbf{Y}(5) = \mathbb{P}(\mathbf{Y} = 5) = .45.$$

Notice that the two random variables $\mathbf{X}$ and $\mathbf{Y}$ are not independent since we have

$$p(2, 5) = \mathbb{P}(\mathbf{X} = 2, \mathbf{Y} = 5) = .13$$
$$\neq \mathbb{P}(\mathbf{X} = 2)\,\mathbb{P}(\mathbf{Y} = 5) = .23 \times .45. = .1035.$$

19.6. Example. Let p be a function defined on $\mathbb{R}^2$ in terms of three real parameters $\alpha \in]0, 1[$, $\beta_i \in]0, 1[$, $i = 1, 2$ by the equation

$$p(x, y) = \begin{cases} \alpha(1 - \beta_1)\beta_1^{y-1} & \text{if } x = 1 \text{ and } y \in \mathbb{N} \\ (1 - \alpha)(1 - \beta_2)\beta_2^{y-1} & \text{if } x = 2 \text{ and } y \in \mathbb{N} \quad (19.9) \\ 0 & \text{in all other cases.} \end{cases}$$

We first show that p is a probability distribution. From (19.9), either p is the product of positive real numbers or vanishes, so $p(x, y) \geq 0$ for all $(x, y) \in \mathbb{R}^2$. Define $\mathcal{S} = \{1, 2\} \times \mathbb{N}$. We have in fact

$$p(x, y) > 0 \quad \Longleftrightarrow \quad (x, y) \in \mathcal{S},$$

and

$$\sum_{(x,y) \in \mathcal{S}} p(x, y) = \sum_{y \in \mathbb{N}} p(1, y) + \sum_{y \in \mathbb{N}} p(2, y)$$
$$= \alpha \sum_{y \in \mathbb{N}} (1 - \beta_1)\beta_1^{y-1} + (1 - \alpha) \sum_{y \in \mathbb{N}} (1 - \beta_2)\beta_2^{y-1}$$
$$= 1$$

because the two formulas $(1 - \beta_i)\beta_i^{y-1}$ $(i = 1, 2)$ in the right-hand side are expressions of two geometric probability distributions with possibly different parameters β_1 and β_2 (see 8.9). As in the preceding example, we define two marginal random variables $\mathbf{X} : (x, y) \mapsto x$

and $\mathbf{Y} : (x,y) \mapsto y$. We have $\mathcal{S}_{\mathbf{X}} = \{1,2\}$ and $\mathcal{S}_{\mathbf{Y}} = \mathbb{N}$. The marginal distributions of $\mathbf{X}$ and $\mathbf{Y}$ are as follows: for all $x, y \in \mathbb{R}$,

$$
\mathbb{P}(\mathbf{X} = x) = \begin{cases} \alpha & \text{if } x = 1, \\ (1 - \alpha) & \text{if } x = 2, \\ 0 & \text{in all other cases ;} \end{cases}
$$

$$
\mathbb{P}(\mathbf{Y} = y) = \begin{cases} \alpha(1 - \beta_1)\beta_1^{y-1} + (1 - \alpha)(1 - \beta_2)\beta_2^{y-1} & \text{if } y \in \mathbb{N}, \\ 0 & \text{otherwise.} \end{cases}
$$

$$(19.10)$$

19.7. Remark. The random variable $\mathbf{Y}$ of Example 19.6 is a 'mixture' of two geometric random variables. Let $\mathbf{X}, \mathbf{X}_1, \dots, \mathbf{X}_n$ be $n+1$ random variables with distribution functions $F_{\mathbf{X}}, F_{\mathbf{X}_1}, \dots, F_{\mathbf{X}_n}$, and let $\alpha_1, \dots, \alpha_n$ be n real numbers, with $0 < \alpha_i < 1$ for $1 \leq i \leq n$ and $\sum_{i=1}^{n} \alpha_i = 1$. The random variable $\mathbf{X}$ is said to be a *mixture* of the random variables $\mathbf{X}_1, \dots, \mathbf{X}_n$ with *mixing parameters* $\alpha_1, \dots, \alpha_n$ if for all real numbers x, we have

$$
F_{\mathbf{X}}(x) = \sum_{i=1}^{n} \alpha_i F_{\mathbf{X}_n}(x). \qquad (19.11)
$$

Clearly, a similar equation holds for probability distributions when the random variables are discrete. Notice that, if the random variables $\mathbf{X}$ and $\mathbf{X}_i$, $1 \leq i \leq n$ are continuous, with respective densities $f_{\mathbf{X}}$ and $f_{\mathbf{X}_i}$, $1 \leq i \leq n$, then $\mathbf{X}$ is a mixture of the random variables $\mathbf{X}_1, \dots, \mathbf{X}_n$ with mixing parameters $\alpha_1, \dots, \alpha_n$ if and only if, for all $x \in \mathbb{R}$, we have

$$
f_{\mathbf{X}}(x) = \sum_{i=1}^{n} \alpha_i f_{\mathbf{X}_i}(x).
$$

Getting back to Example 19.6, the marginal random variable $\mathbf{Y}$ is a mixture of two random discrete variables $\mathbf{Y}_i$ ($i = 1, 2$), both having a geometric distribution with parameter β_i, that is,

$$
\mathbb{P}(\mathbf{Y}_i = y) = \begin{cases} (1 - \beta_i)\beta_i^{y-1} & \text{if } y \in \mathbb{N}, \\ 0 & \text{otherwise.} \end{cases}
$$

The mixing parameters are α and $1 - \alpha$. In view of Eq. (19.10), we have indeed, for all $y \in \mathbb{R}$,

$$
\mathbb{P}(\mathbf{Y} = y) = \alpha \mathbb{P}(\mathbf{Y}_1 = y) + (1 - \alpha)\mathbb{P}(\mathbf{Y}_2 = y).
$$

The General Case

19.8. Definition. Let $\mathcal{X} = (\mathbf{X}_i)_{1 \leq i \leq n}$ be a finite vector of random variables on the same sample space $\mathbb{R}^n$. The *multivariate distribution function* of $\mathcal{X}$ is the function $F_{\mathcal{X}} : \mathbb{R}^n \to [0, 1]$ defined by

$$F_{\mathcal{X}}(x_1, \ldots, x_n) = \mathbb{P}(\mathbf{X}_1 \leq x_1, \ldots, \mathbf{X}_n \leq x_n). \tag{19.12}$$

This generalizes Def. 19.4.

In the case of only two jointly distributed random variables $\mathbf{X}$ and $\mathbf{Y}$, we will sometimes adopt a more explicit notation and write

$$F_{\mathbf{X}, \mathbf{Y}}(x, y) = \mathbb{P}(\mathbf{X} \leq x, \mathbf{Y} \leq y)$$

for the *bivariate distribution function* of the pair of random variables $\mathcal{X} = (\mathbf{X}, \mathbf{Y})$. Notice that $\mathbb{P}(\mathbf{X} \leq x, \mathbf{Y} \leq y)$ is the probability assigned to a subset of $\mathbb{R}^2$ forming an infinite half-closed 'rectangle' defined by the intersection of two half-planes, namely,

$$\{(s, t) \in \mathbb{R}^2 \mid s \leq x\} \cap \{(s, t) \in \mathbb{R}^2 \mid t \leq y\}. \tag{19.13}$$

In the sequel, we only deal with this binary case. (The extension of the concepts to a family of n jointly distributed random variables is, in most cases, straightforward; see, for example, Problem 8.)

Make note of the following important identity which is illustrated by Fig. 19.1.

19.9. Theorem. *Let $F_{\mathbf{X}, \mathbf{Y}}$ be the bivariate distribution function of two r.v.'s $\mathbf{X}$ and $\mathbf{Y}$. For any real numbers $a < b$ and $c < d$, we have*

$$\mathbb{P}(a < \mathbf{X} \leq b, c < \mathbf{Y} \leq d)$$
$$= F_{\mathbf{X}, \mathbf{Y}}(b, d) + F_{\mathbf{X}, \mathbf{Y}}(a, c) - F_{\mathbf{X}, \mathbf{Y}}(a, d) - F_{\mathbf{X}, \mathbf{Y}}(b, c).$$

We leave the proof as Problem 7.

In the spirit of Def. 19.4 for the discrete case, the marginal distribution function of the random variable $\mathbf{X}$ can be defined from the bivariate distribution function $F_{\mathbf{X}, \mathbf{Y}}$ by accumulating the mass pertaining to the other dimension, and similarly for the marginal distribution function of $\mathbf{Y}$. However, since we are dealing with distribution functions, we take limits rather than sums.

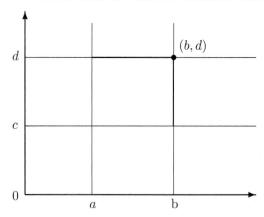

Figure 19.1: Illustration of Theorem 19.9.

19.10. Definition. Let $F_{\mathbf{X},\mathbf{Y}}$ be the bivariate distribution function of the pair of r.v.'s $(\mathbf{X},\mathbf{Y})$; then, the function $F_{\mathbf{X}} : \mathbb{R} \to [0,1]$ defined by

$$F_{\mathbf{X}}(x) = \mathbb{P}(\mathbf{X} \le x) = \lim_{y \to \infty} F_{\mathbf{X},\mathbf{Y}}(x,y)$$

is the *marginal distribution function* of $\mathbf{X}$ in the pair $(\mathbf{X},\mathbf{Y})$. Thus, in the situation of Fig. 19.1, $F_{\mathbf{X}}(b)$ is the probability assigned to the closed half-plane to the left of the line $x = b$, and including also that line. Similarly,

$$F_{\mathbf{Y}}(y) = \mathbb{P}(\mathbf{Y} \le y) = \lim_{x \to \infty} F_{\mathbf{X},\mathbf{Y}}(x,y)$$

defines *marginal distribution function* of $\mathbf{Y}$ in the pair $(\mathbf{X},\mathbf{Y})$.

19.11. Definition. Let $\mathbf{X}$ and $\mathbf{Y}$ be two jointly distributed random variables, with bivariate distribution function $F_{\mathbf{X},\mathbf{Y}}$. Suppose that the following derivative exists:

$$f_{\mathbf{X},\mathbf{Y}}(x,y) = \frac{\partial^2}{\partial x\, \partial y} F_{\mathbf{X},\mathbf{Y}}(x,y). \tag{19.14}$$

Then, the function $f_{\mathbf{X},\mathbf{Y}} : \mathbb{R} \to\,]0,\infty[$ defined by (19.14) is called the *bivariate density (function)* of $(\mathbf{X},\mathbf{Y})$. In the case of a finite vector $\mathcal{X} = (\mathbf{X}_1,\ldots,\mathbf{X}_n)$, with $n > 2$, the corresponding function $f_{\mathcal{X}} : \mathbb{R}^n \to\,]0,\infty[$ is referred to as the *multivariate density* of $\mathcal{X}$.

   If the function $f_{\mathbf{X},\mathbf{Y}}$ is continuous, then the two random variables $\mathbf{X}$ and $\mathbf{Y}$ are continuous in the sense of 14.1. This means that their density functions are defined. In this context, these densities are referred to as the *marginal densities* of $\mathbf{X}$ and $\mathbf{Y}$ in the pair $(\mathbf{X}, \mathbf{Y})$. They are obtained from

$$f_{\mathbf{X}}(x) = \int_{-\infty}^{\infty} f_{\mathbf{X},\mathbf{Y}}(x,y)dy$$
$$f_{\mathbf{Y}}(y) = \int_{-\infty}^{\infty} f_{\mathbf{X},\mathbf{Y}}(x,y)dx. \tag{19.15}$$

19.12. Example. A bivariate distribution function is said to be *exponential* if its marginals are exponential distribution functions (Gumbel, 1960, Johnson and Kotz, 1972, Kotz et al., 1985). Two examples of such distribution functions are, for $x \geq 0$ and $y \geq 0$,

$$F_{\mathbf{X},\mathbf{Y}}(x,y) = 1 - e^{-x} - e^{-y} + e^{-x-y-\delta xy} \qquad (0 \leq \delta \leq 1);$$
$$F_{\mathbf{X},\mathbf{Y}}(x,y) = (1 - e^{-x})(1 - e^{-y})(1 + \alpha e^{-x-y}) \qquad (-1 \leq \alpha \leq 1),$$

with $F_{\mathbf{X},\mathbf{Y}} = 0$ if $x < 0$ or $y < 0$. It is easily seen that, in both cases, the marginals are exponential with a parameter equal to 1.

19.13. Example. Another example of a bivariate exponential was proposed by Marshall and Olkin (1967). The distribution function $F_{\mathbf{X},\mathbf{Y}}$ is defined implicitly, for $x, y \geq 0$, by

$$\mathbb{P}(\mathbf{X} > x, \mathbf{Y} > y) = \exp(-\lambda_1 x - \lambda_2 y - \lambda_{12} \max\{x, y\}) \tag{19.16}$$

where λ_1, λ_2, and λ_{12} are nonnegative, with $\lambda_1 + \lambda_{12} > 0$ and $\lambda_2 + \lambda_{12} > 0$. This particular distribution arises in the modeling of the 'fatal shock' phenomenon. Consider a tool made of two critical parts, labeled '1' and '2.' Each of these parts may be disabled by a fatal shock. The occurrence of these fatal shocks are governed by two independent Poisson processes with parameters λ_1 and λ_2 (see 24.10). If $\mathbf{X}$ and $\mathbf{Y}$ measure the life span of each of these two parts, then (19.16) specifies their joint distribution function.[2]

Chapter 20 is dedicated to the important example of the bivariate normal distribution. To end this section, we recast the concept of independence of two random variables (cf. 12.11) in the framework of bivariate distributions.

---

[2]This equation also arises from the following mechanism. Suppose that $\mathbf{X} = \min\{\mathbf{V}, \mathbf{W}\}$ and $\mathbf{Y} = \min\{\mathbf{U}, \mathbf{W}\}$, where $\mathbf{U}$, $\mathbf{V}$, and $\mathbf{W}$ are exponentially distributed independent r.v.'s with parameters λ_1, λ_2, and λ_{12}, respectively. Equation (19.16) follows.

19.14. Theorem. *Two random variables* $\mathbf{X}$ *and* $\mathbf{Y}$ *with bivariate distribution function* $F_{\mathbf{X},\mathbf{Y}}$ *are independent if and only if*

$$F_{\mathbf{X},\mathbf{Y}}(x,y) = F_{\mathbf{X}}(x)F_{\mathbf{Y}}(y), \qquad (x,y \in \mathbb{R}), \tag{19.17}$$

or equivalently, if the bivariate density exists,

$$f_{\mathbf{X},\mathbf{Y}}(x,y) = f_{\mathbf{X}}(x)f_{\mathbf{Y}}(y), \qquad (x,y \in \mathbb{R}). \tag{19.18}$$

Equation (19.17) follows immediately from Def. 12.11 (specifically, Eq. (12.12)) and (19.18) obtains by differentiating (19.17) with respect to x and y.

Bivariate Moment Generating Function

We also mention that the concept of moment generating function introduced in 18.1 can be extended to the multivariate situation. In the definition below, we restrict consideration to the bivariate case.

19.15. Definition. The *(joint) moment generating function*—or *(joint) M.G.F.*—of a pair $\mathcal{X} = (\mathbf{X}_1, \mathbf{X}_2)$ of two jointly distributed random variables with distribution function $F_{\mathcal{X}}$ is the function

$$M_{\mathcal{X}}(\theta_1, \theta_2) = E\left(e^{\theta_1 \mathbf{X}_1 + \theta_2 \mathbf{X}_2}\right) \tag{19.19}$$

$$= \int_{-\infty}^{\infty} \int_{-\infty}^{\infty} e^{\theta_1 x + \theta_2 y} dF_{\mathcal{X}}(x_1, x_2) \tag{19.20}$$

which is assumed to be defined for all ordered pairs of real numbers (θ_1, θ_2) in an open subset of $\mathbb{R}^2$ containing $(0,0)$. As in the univariate case, the joint distribution function is completely specified by the M.G.F. when the latter is defined.[3] Moreover, as indicated by the next theorem, the M.G.F. can be used to compute the moments of the marginal distributions of $\mathbf{X}_1$ and $\mathbf{X}_2$ as well as the joint moments such as the covariance. We leave the proof as Problem 12.

19.16. Theorem. *Let* $\mathcal{X} = (\mathbf{X}_1, \mathbf{X}_2)$ *be a pair of random variables with joint distribution function* $F_{\mathcal{X}}$ *and a M.G.F. denoted by* $M_{\mathcal{X}}(\theta_1, \theta_2)$; *then, the* nth *partial derivative of the function* $M_{\mathcal{X}}$ *with respect to* θ_i, *for* $i = 1, 2$ *has the form*

$$\frac{\partial^n M_{\mathcal{X}}}{\partial \theta_i} = \int_{-\infty}^{\infty} \int_{-\infty}^{\infty} x_i^n \, e^{\theta_1 x_1 + \theta_2 x_2} \, dF_{\mathcal{X}}(x_1, x_2). \tag{19.21}$$

---

[3]The closely related complex valued 'joint characteristic function' plays a similar role and is always defined (cf. Remark 18.11).

Accordingly, for $i = 1, 2$,

$$E(\mathbf{X}_i^n) = \frac{\partial^n M_\mathcal{X}}{\partial \theta_i}\Big|_{\theta_1 = \theta_2 = 0}. \tag{19.22}$$

We also have, for the first joint moment,

$$\frac{\partial^2 M_\mathcal{X}}{\partial \theta_1 \partial \theta_2} = \int_{-\infty}^{\infty} \int_{-\infty}^{\infty} x_1 x_2 e^{\theta_1 x_1 + \theta_2 x_2} dF_\mathcal{X}(x_1, x_2), \tag{19.23}$$

yielding

$$E(\mathbf{X}_1 \mathbf{X}_2) = \frac{\partial^2 M_\mathcal{X}}{\partial \theta_1 \partial \theta_2}\Big|_{\theta_1 = \theta_2 = 0}. \tag{19.24}$$

Conditional Distributions

If $\mathbf{X}$ and $\mathbf{Y}$ are jointly distributed random variables, the 'conditional distribution function of $\mathbf{X}$ given the event $\mathbf{Y} = y$' may be defined, if $\mathbb{P}(\mathbf{Y} = y) > 0$, by

$$F_{\mathbf{X}|\mathbf{Y}}(x|y) = \mathbb{P}(\mathbf{X} \le x \mid \mathbf{Y} = y). \tag{19.25}$$

Such a definition is not helpful when $\mathbf{Y}$ is continuous because the event $\{y \in \mathbb{R} \mid \mathbb{P}(\mathbf{Y} = y) > 0\}$ has probability 0. The situation is more promising when the bivariate density of $\mathbf{X}$ and $\mathbf{Y}$ is defined. We combine both cases in the next definition.

19.17. Definition. Suppose that $\mathbf{X}$ and $\mathbf{Y}$ have a joint density $f_{\mathbf{X},\mathbf{Y}}$ and that the marginal density of $\mathbf{Y}$ satisfies $f_{\mathbf{Y}}(y) > 0$ for a particular $y \in \mathbb{R}$. The *conditional density of $\mathbf{X}$ given $\mathbf{Y} = y$* is defined by the equation

$$f_{\mathbf{X}|\mathbf{Y}}(x|y) = \frac{f_{\mathbf{X},\mathbf{Y}}(x, y)}{f_{\mathbf{Y}}(y)}. \tag{19.26}$$

We can then use $f_{\mathbf{X}|\mathbf{Y}}$ to obtain $F_{\mathbf{X}|\mathbf{Y}}$ in this case. Summing up, we define the *conditional distribution function of $\mathbf{X}$ given $\mathbf{Y} = y$* as

$$F_{\mathbf{X}|\mathbf{Y}}(x|y) = \begin{cases} \mathbb{P}(\mathbf{X} \le x \mid \mathbf{Y} = y) & \text{if } \mathbb{P}(\mathbf{Y} = y) > 0, \\ \int_{-\infty}^{x} f_{\mathbf{X}|\mathbf{Y}}(z|y) \, dz & \text{if } f_{\mathbf{X},\mathbf{Y}} \text{ exists and } f_{\mathbf{Y}}(y) > 0. \end{cases}$$

If $G : \mathbb{R} \to \mathbb{R}$ is a continuous function, the *conditional expectation* of $G(\mathbf{X})$ *given* $\mathbf{Y} = y$ is defined—if $f_{\mathbf{X},\mathbf{Y}}$ exists and $f_{\mathbf{Y}}(y) > 0$—by

$$E\left(G(\mathbf{X}) \mid \mathbf{Y} = y\right) = \int_{-\infty}^{\infty} G(x)\, dF_{\mathbf{X}|\mathbf{Y}}(x|y) \qquad (19.27)$$

$$= \int_{-\infty}^{\infty} G(x) f_{\mathbf{X}|\mathbf{Y}}(x|y)\, dx \qquad (19.28)$$

which is a conditional version of Eq. (15.3). In view of (19.25), Eq. (19.27) is of course valid even if $f_{\mathbf{X},\mathbf{Y}}$ does not exist when $\mathbb{P}(\mathbf{Y} = y) > 0$. The *conditional variance* of $\mathbf{X}$ given $\mathbf{Y} = y$ is

$$Var(\mathbf{X} \mid \mathbf{Y} = y) = E(\mathbf{X}^2 \mid \mathbf{Y} = y) - E(\mathbf{X} \mid \mathbf{Y} = y)^2 \qquad (19.29)$$

when the expectations in the r.h.s. are defined.

19.18. Example. For the r.v.'s $\mathbf{X}$ and $\mathbf{Y}$ of Example 19.5, whose bivariate probability distribution is defined by Table 19.1, we have

$$E(\mathbf{X} \mid \mathbf{Y} = 6) = \frac{1 \times .10 + 2 \times .05 + 3 \times .15}{.30} \approx 2.17\,,$$

and

$$Var(\mathbf{Y} \mid \mathbf{X} = 3) = E(\mathbf{Y}^2 \mid \mathbf{X} = 3) - E(\mathbf{Y} \mid \mathbf{X} = 3)^2$$
$$= 28 - 27.5625 = .4375\,.$$

19.19. Example. In Example 19.6, we also have two random variables $\mathbf{X}$ and $\mathbf{Y}$, whose bivariate distribution is specified by (19.9). The conditional expectation $E(\mathbf{Y} \mid \mathbf{X} = 1)$ and conditional variance $Var(\mathbf{Y} \mid \mathbf{X} = 1)$ are the expectation and the variance of a geometric random variable with parameter β_1, namely (cf. 8.9 and Table 15.1)

$$E(\mathbf{Y} \mid \mathbf{X} = 1) = \frac{1}{\beta_1}, \qquad Var(\mathbf{Y} \mid \mathbf{X} = 1) = \frac{\beta_1}{1 - \beta_1^2}\,.$$

19.20. Remark. We can attribute a meaning to expressions such as $E\left(G(\mathbf{X}) \mid \mathbf{Y}\right)$ (again, with $G : \mathbb{R} \to \mathbb{R}$ a continuous function). We assume in this remark and in the rest of this chapter that all expectations are finite. Implicitly, the conditional expectation specified by (19.27) or (19.28) can be regarded as defining a function

$$\mathbf{Z} : \Omega \xrightarrow{\mathbf{Y}} \mathbb{R} \xrightarrow{E(G(\mathbf{X})|\,.\,)} \mathbb{R}$$

mapping the sample space into the reals. (Thus, if $\mathbf{Y} = y$, then $E\left(G(\mathbf{X}) \mid \mathbf{Y}\right)$ takes the value $E\left(G(\mathbf{X}) \mid \mathbf{Y} = y\right)$.) It can in fact be shown—we shall not do this here—that $\mathbf{Z}$ is a random variable. Taking G to be the identity function, we shall prove here that

$$E(\mathbf{X}) = E\left(E(\mathbf{X} \mid \mathbf{Y})\right). \tag{19.30}$$

Similar remarks apply to the conditional variance. Equation (19.29) leads to

$$Var(\mathbf{X} \mid \mathbf{Y}) = E(\mathbf{X}^2 \mid \mathbf{Y}) - E(\mathbf{X} \mid \mathbf{Y})^2,$$

an equation relating three random variables. Taking expectations on both sides gives

$$E(Var(\mathbf{X} \mid \mathbf{Y})) = E(\mathbf{X}^2) - E(E(\mathbf{X} \mid \mathbf{Y})^2) \tag{19.31}$$

because, as an application of (19.30), we have $E(\mathbf{X}^2) = E\left(E(\mathbf{X}^2 \mid \mathbf{Y})\right)$. We now subtract $E(\mathbf{X})^2$ on both sides of (19.31). Gathering terms, using (19.30), and rearranging finally gives

$$Var(\mathbf{X}) = E(Var(\mathbf{X} \mid \mathbf{Y})) + Var(E(\mathbf{X} \mid \mathbf{Y})), \tag{19.32}$$

an equation slightly more complicated than (19.30) but having the neat wording: THE VARIANCE IS EQUAL TO THE EXPECTATION OF THE CONDITIONAL VARIANCE PLUS THE VARIANCE OF THE CONDITONAL EXPECTATION.

19.21. Warning. The double expectation in (19.30) or (19.31) implements a rather compact notation. The student should be mindful of the fact that, in a formula such as $E(E(G(\mathbf{X}, \mathbf{Y})) \mid \mathbf{Y})$, for instance, (with $\mathbf{X}$ and $\mathbf{Y}$ two r.v.'s and G a continuous function), the first expectation applies to the second instance of the r.v. $\mathbf{Y}$, so that

$$E(E(G(\mathbf{X}, \mathbf{Y})) \mid \mathbf{Y}) = E(G(\mathbf{X}, \mathbf{Y})),$$

which must be regarded as a special case of (19.30).

We end up this chapter by proving (19.30) and (19.32). The next theorem is a step in that direction.

19.22. Theorem. *For any two random variables* $\mathbf{X}$ *and* $\mathbf{Y}$ *and any continuous function* $G : \mathbb{R} \to \mathbb{R}$*, we have*

$$E\left(G(\mathbf{Y})\mathbf{X}\right) = E\left(G(\mathbf{Y})E(\mathbf{X} \mid \mathbf{Y})\right). \tag{19.33}$$

PROOF. We only give a proof for the discrete case. We denote by $p_{\mathbf{X},\mathbf{Y}}$ the joint probability distribution of the random variables $\mathbf{X}$ and $\mathbf{Y}$, and by $p_{\mathbf{X}}$ and $p_{\mathbf{Y}}$ their marginal probability distributions. (Thus, $p_{\mathbf{X}}(x) = \mathbb{P}(\mathbf{X} = x)$ and $p_{\mathbf{Y}}(y) = \mathbb{P}(\mathbf{Y} = y)$ for all $x, y \in \mathbb{R}$.) As before, we write

$$S_{\mathbf{X}} = \{x \in \mathbb{R} \mid p_{\mathbf{X}}(x) > 0\}, \quad S_{\mathbf{Y}} = \{y \in \mathbb{R} \mid p_{\mathbf{Y}}(y) > 0\}.$$

(These sets are at most countable since $\mathbf{X}$ and $\mathbf{Y}$ are discrete.) Both $G(\mathbf{Y})$ and $E(\mathbf{X} \mid \mathbf{Y})$ are random variables, which take values $G(y)$ and $E(\mathbf{X} \mid \mathbf{Y} = y)$, respectively, when $\mathbf{Y} = y$. We have thus successively

$$E\left(G(\mathbf{Y})E(\mathbf{X} \mid \mathbf{Y})\right) = \sum_{y \in S_{\mathbf{Y}}} G(y)\, E(\mathbf{X} \mid \mathbf{Y} = y)\, p_{\mathbf{Y}}(y)$$

$$= \sum_{y \in S_{\mathbf{Y}}} G(y) \left(\sum_{x \in S_{\mathbf{X}}} x\, \frac{p_{\mathbf{X},\mathbf{Y}}(x,y)}{p_{\mathbf{Y}}(y)} \right) p_{\mathbf{Y}}(y)$$

$$= \sum_{x \in S_{\mathbf{X}}} \sum_{y \in S_{\mathbf{Y}}} G(y)\, x\, p_{\mathbf{X},\mathbf{Y}}(x,y)$$

$$= E\left(G(\mathbf{Y})\mathbf{X}\right),$$

as asserted.   □

19.23. Corollary.

$$E(\mathbf{X}) = E\left(E(\mathbf{X} \mid \mathbf{Y})\right).$$

This follows by setting $G(\mathbf{Y}) = 1$ in (19.33).

19.24. Theorem. *For any two random variables* $\mathbf{X}$ *and* $\mathbf{Y}$

$$Var(\mathbf{X}) = E(Var(\mathbf{X} \mid \mathbf{Y})) + Var(E(\mathbf{X} \mid \mathbf{Y})). \tag{19.34}$$

PROOF. It is easily verified that, for any random variable $\mathbf{Z}$ and constant α, we have

$$E\left((\mathbf{Z} - \alpha)^2\right) = E\left((\mathbf{Z}^2 - E(\mathbf{Z})^2) + (E(\mathbf{Z}) - \alpha)\right)^2 \tag{19.35}$$

(cf. Problem 15). Setting $\mathbf{Z} = \mathbf{X}$, Eq. (19.35) applies in particular when all the expectations in (19.35) are conditionalized by the random variable $\mathbf{Y}$. So, with $\alpha = E(\mathbf{X})$, we obtain

$$E[(\mathbf{X} - E(\mathbf{X}))^2 \mid \mathbf{Y}] = E[(\mathbf{X} - E(\mathbf{X} \mid \mathbf{Y})^2 \mid \mathbf{Y}] + [E(\mathbf{X} \mid \mathbf{Y}) - E(\mathbf{X}))]^2.$$

Taking expectations on both sides of this equation and remembering our warning in 19.21, we get (19.34).   □

Problems

1. Argue that the functions defined by (19.4) are genuine random variables.

2. Prove Theorem 19.2 in the case $\mathbb{P}(\mathbf{X}_i = m) = 0$.

3. In 19.1, the event described by '$\mathbf{X}_i = m_i$' is the disjoint union of all the (punctual) events $\{m_1, \ldots, m_i, \ldots, m_k\}$ with $\sum_{j \neq i} m_j = n - m_i$. Prove Theorem 19.2 by summing the probabilities of all those events.

4. Prove Theorem 19.3. (Hint: Extend the proof of Theorem 9.8.)

5. Verify that the marginal distributions of the random variables $\mathbf{X}$ and $\mathbf{Y}$ are as stated in 19.6. Could these two random variable be independent for some values of the parameters?

6. Suppose that $\mathbf{X}$ is a mixture of the two random variables $\mathbf{Y}$ and $\mathbf{Z}$ with mixture parameters α and $1 - \alpha$ in the sense of Remark 19.7. Find expressions for the mean and the variance of $\mathbf{X}$ in terms of the means and the variances of $\mathbf{Y}$ and $\mathbf{Z}$. What happens when the expectations of $\mathbf{Y}$ and $\mathbf{Z}$ coincide?

7. Prove Theorem 19.9.

8. Define the marginal distribution function of the r.v. $\mathbf{X}_i$ in the collection $\mathcal{X} = \{\mathbf{X}_1, \ldots, \mathbf{X}_n\}$. The random variables are not assumed to be discrete.

9. In the context of Examples 19.12 and 19.13, check that the marginal distributions of the three distribution functions given are indeed exponentials and propose other examples of bivariate exponential distributions.

10. Suppose that the bivariate distribution function $F_{\mathbf{X}, \mathbf{Y}}(x, y)$ of two independent random variables $\mathbf{X}$ and $\mathbf{Y}$ vanishes for $x < 0$ or $y < 0$ (or both) and also satisfies the equation $F_{\mathbf{X}, \mathbf{Y}}(x, y) = h(x + y)$ for some unknown function h. What can you say about the form of $F_{\mathbf{X}, \mathbf{Y}}$ and of the marginal distributions? (Hint: use Lemma 14.8).

11. For the random variables $\mathbf{X}$ and $\mathbf{Y}$ of Example 19.6, compute the conditional expectation and the conditional variance of the random variable $\mathbf{X}$, the conditional event being '$\mathbf{Y} = 17$.'

12. Prove Theorem 19.16.

13. Prove that if the multivariate distribution of a vector $\mathcal{X} = (\mathbf{X}_1, \ldots, \mathbf{X}_n)$ is given, then the multivariate distribution of any subvector $(\mathbf{X}_{n_1}, \ldots, \mathbf{X}_{n_k})$ of $\mathcal{X}$ is thereby also specified. (Hint: Use the joint M.G.F.)

14. Prove Theorem 15.4 in the continuous case. (Thus, you may assume that the joint density exists.)

15. Verify that (19.35) holds for any random variable $\mathbf{Z}$ and constant α.

Chapter 20

Bivariate Normal Distributions

Density Distribution

20.1. Definition. A pair $\mathcal{X} = (\mathbf{X}_1, \mathbf{X}_2)$ of jointly distributed random variables is *normal* or has a *normal distribution* if its joint density has the form

$$f_{\mathcal{X}}(x_1, x_2) = \frac{1}{2\pi\sigma_1\sigma_2\sqrt{1 - \rho^2}} \times$$

$$\exp\left\{\frac{-1}{2(1 - \rho^2)}\left[\left(\frac{x_1 - \mu_1}{\sigma_1}\right)^2 + \left(\frac{x_2 - \mu_2}{\sigma_2}\right)^2 - 2\rho\left(\frac{x_1 - \mu_1}{\sigma_1}\right)\left(\frac{x_2 - \mu_2}{\sigma_2}\right)\right]\right\},$$
$$(20.1)$$

in which μ_1, μ_2, σ_1, $\sigma_2 > 0$, and $0 < \rho < 1$ are real parameters. The interpretation of some of these parameters is suggested by the notation and made clear by Theorem 20.2 and Corollary 20.4.

20.2. Theorem. *If a pair $\mathcal{X} = (\mathbf{X}_1, \mathbf{X}_2)$ of random variables has a normal joint density of the form (20.1), then the marginal densities f_i $(i = 1, 2)$ of $\mathbf{X}_1$ and $\mathbf{X}_2$ are also normal, and we have*

$$f_i(x_i) = \frac{1}{\sigma_i\sqrt{2\pi}}e^{-\frac{1}{2}\left(\frac{x_i - \mu_i}{\sigma_i}\right)^2}, \qquad (i = 1, 2, \ x_i \in \mathbb{R}). \quad (20.2)$$

Thus, the parameters μ_i and σ_i $(i = 1, 2)$ in (20.1) are, respectively, the means and the standard deviations of the marginal distributions of $\mathbf{X}_1$ and $\mathbf{X}_2$.

PROOF. Adding and subtracting $(x_1 - \mu_1)^2/(2\sigma_1^2)$ in the exponent of (20.1) and integrating over x_2—cf. (19.15)—we obtain, after gathering the terms appropriately,

$$f_1(x_1) = \int_{-\infty}^{\infty} f_{\mathcal{X}}(x_1, x_2)\, dx_2 = \frac{1}{\sigma_1\sqrt{2\pi}}\, e^{-\frac{1}{2}\left(\frac{x_1-\mu_1}{\sigma_1}\right)^2} \times A,$$

with

$$A = \frac{1}{\sigma_2\sqrt{(1-\rho^2)2\pi}} \int_{-\infty}^{\infty} \exp\left(-\frac{1}{2}\left(\frac{x_2 - (\mu_2 + \frac{\sigma_2}{\sigma_1}\rho(x_1 - \mu_1))}{\sigma_2\sqrt{1-\rho^2}}\right)^2\right) dx_2.$$

The factor A is the integral over $\mathbb{R}$ of the density function of a normal r.v. with an expectation equal to $\mu_2 + \frac{\sigma_2}{\sigma_1}\rho(x_1 - \mu_1)$ and a variance equal to $\sigma_2^2(1 - \rho^2)$. (See Chapter 15, and in particular the last line of Table 15.1.) Accordingly, $A = 1$ and (20.2) obtains with $i = 1$. The case $i = 2$ follows by symmetry. The facts that the parameters μ_i and σ_i $(i = 1, 2)$ are, respectively, the means and the standard deviations of the random variables $\mathbf{X}_1$ and $\mathbf{X}_2$ are clear from the form of (20.2). □

Joint Moment Generating Function

The normality of jointly distributed r.v.'s can also be specified by their joint moment generating function (cf. Def. 19.15).

20.3. Theorem. *The joint distribution of a pair $\mathcal{X} = (\mathbf{X}_1, \mathbf{X}_2)$ of random variables is normal if and only if its M.G.F. exists and satisfies the equation*

$$M_{\mathbf{X}_1, \mathbf{X}_2}(\theta_1, \theta_2) =$$
$$\exp\left(\theta_1\mu_1 + \theta_2\mu_2 + \frac{1}{2}(\theta_1^2\sigma_1^2 + \theta_2^2\sigma_2^2 + 2\theta_1\theta_2\sigma_1\sigma_2\rho)\right) \qquad (20.3)$$

with $\mu_1, \mu_2 \in \mathbb{R}$, $\sigma_1 > 0$, $\sigma_2 > 0$, $0 < \rho < 1$, the parameters of the bivariate density in (20.1), and $M_{\mathbf{X}_1, \mathbf{X}_2}(\theta_1, \theta_2)$ is defined in an open interval containing $(0, 0)$.

To get a proof of this result, we apply Def. 19.15 and compute the integral

$$E\left(e^{\theta_1 \mathbf{X}_1 + \theta_2 \mathbf{X}_2}\right) = \int_{-\infty}^{\infty}\int_{-\infty}^{\infty} e^{\theta_1 x_1 + \theta_2 x_2} f_{\mathcal{X}}(x_1, x_2)\, dx_1\, dx_2.$$

We leave this task to the student (see Problem 1). We mention two applications of (20.3).

20.4. Corollary. *The parameter ρ in (20.1) and (20.3) is the correlation coefficient of the random variables $\mathbf{X}_1$ and $\mathbf{X}_2$.*

PROOF. Applying (19.24) to the M.G.F. defined by (20.3) gives

$$E(\mathbf{X}_1\mathbf{X}_2) = \frac{\partial^2 M_{\mathcal{X}}}{\partial\theta_1\partial\theta_2}\Big|_{\theta_1=\theta_2=0} = \mu_1\mu_2 + \sigma_1\sigma_2\rho,$$

and so $\rho = \frac{E(\mathbf{X}_1\mathbf{X}_2)-\mu_1\mu_2}{\sigma_1\sigma_2}$, establishing the corollary. $\qquad\square$

20.5. Theorem. *If $\mathbf{X}$ and $\mathbf{Y}$ are jointly normal, then for any real numbers α_1, α_2, β_1, and β_2, with at least one α_i $(i = 1, 2)$ and one β_i $(i = 1, 2)$ not equal to zero, the random variables $\alpha_1\mathbf{X}+\alpha_2\mathbf{Y}$ and $\beta_1\mathbf{X} + \beta_2\mathbf{Y}$ are also jointly normal.*

(The proof is left as Problem 4.)

Independence and Correlation

We now go back to a result announced earlier (cf. Theorem 16.2) and give a complete proof.

20.6. Theorem. *Two jointly and normally distributed random variables $\mathbf{X}_1$ and $\mathbf{X}_2$ are independent if and only if their covariance vanishes.*

PROOF. Let $\mathcal{X} = (\mathbf{X}_1, \mathbf{X}_2)$ be the vector of these random variables and let f_i, $i = 1, 2$, be their marginal densities. If they are independent, the joint density $f_{\mathcal{X}}$ satisfies, by (19.18) and (14.12),

$$
\begin{aligned}
f_{\mathcal{X}}(x_1, x_2) &= f_1(x_1)f_2(x_2) \\
&= \frac{1}{\sigma_1\sqrt{2\pi}}e^{-\frac{1}{2}\left(\frac{x_1-\mu_1}{\sigma_1}\right)^2}\frac{1}{\sigma_2\sqrt{2\pi}}e^{-\frac{1}{2}\left(\frac{x_2-\mu_2}{\sigma_2}\right)^2} \quad (20.4) \\
&= \frac{1}{2\pi\sigma_1\sigma_2}e^{-\frac{1}{2}\left(\left(\frac{x_1-\mu_1}{\sigma_1}\right)^2+\left(\frac{x_2-\mu_2}{\sigma_2}\right)^2\right)}. \quad (20.5)
\end{aligned}
$$

The r.h.s. of (20.5) must be equal to the r.h.s. of (20.1). It is easily seen (cf. Problem 2) that this can happen only if $\rho = 0$ in (20.1), and so $Cov(\mathbf{X}_1, \mathbf{X}_2) = 0$.

Conversely, if $Cov(\mathbf{X}_1, \mathbf{X}_2) = 0$, then $\rho = 0$ in (20.1), which can then be put in the form of (20.4). The conclusion follows from Theorem 19.14.   □

20.7. Corollary. *If* $\mathbf{X}$ *and* $\mathbf{Y}$ *are jointly normally distributed and have the same variance, then* $\mathbf{X} + \mathbf{Y}$ *and* $\mathbf{X} - \mathbf{Y}$ *are independent.*

PROOF.   By Theorem 20.5, the random variables $\mathbf{X} + \mathbf{Y}$ and $\mathbf{X} - \mathbf{Y}$ are jointly normally distributed. In view of Theorem 20.6, it suffices thus to show that the covariance of $\mathbf{X} + \mathbf{Y}$ and $\mathbf{X} - \mathbf{Y}$ vanishes, which is the case. Indeed, we have successively

$$
\begin{aligned}
Cov(\mathbf{X} + \mathbf{Y}, \mathbf{X} - \mathbf{Y}) &= E((\mathbf{X} + \mathbf{Y})(\mathbf{X} - \mathbf{Y})) - E(\mathbf{X} + \mathbf{Y})E(\mathbf{X} - \mathbf{Y}) \\
&= \left(E(\mathbf{X}^2) - E(\mathbf{X})^2\right) - \left(E(\mathbf{Y}^2) - E(\mathbf{Y})^2\right) \\
&= 0
\end{aligned}
$$

because $\mathbf{X}$ and $\mathbf{Y}$ have the same variance.   □

Conditional Density and Expectation

20.8. Theorem. *Let* $\mathcal{X} = (\mathbf{X}_1, \mathbf{X}_2)$ *be a pair of random variables with a normal joint density function* $f_{\mathcal{X}}$ *as specified by* (20.1). *The conditional density of* $\mathbf{X}_1$ *given that* $\mathbf{X}_2 = x_2$ *is defined for all* $(x_1, x_2) \in \mathbb{R}^2$ *by*

$$
f_{\mathbf{X}_1 | \mathbf{X}_2}(x_1 | x_2) = \frac{f_{\mathcal{X}}(x_1, x_2)}{f_{\mathbf{X}_2}(x_2)} \tag{20.6}
$$

$$
= \frac{1}{\sigma_1 \sqrt{(1 - \rho^2)2\pi}} \exp\left(-\frac{1}{2}\left(\frac{x_1 - (\mu_1 + \frac{\sigma_1}{\sigma_2}\rho(x_2 - \mu_2))}{\sigma_1\sqrt{1 - \rho^2}}\right)^2\right). \tag{20.7}
$$

As indicated by (20.6), the conditional density of $\mathbf{X}_1$ is obtained by dividing the normal joint density of $(\mathbf{X}_1, \mathbf{X}_2)$ by the marginal density of $\mathbf{X}_2$—applying, thus, (20.1) and (20.2). Examining the r.h.s. of (20.7), it is clear that it specifies the density of a normal random variable with expectation and variance given by the two equations:

$$
E(\mathbf{X}_1 | \mathbf{X}_2 = x_2) = \mu_1 + \frac{\sigma_1}{\sigma_2}\rho(x_2 - \mu_2), \tag{20.8}
$$

$$
Var(\mathbf{X}_1 | \mathbf{X}_2 = x_2) = \sigma_1^2(1 - \rho^2). \tag{20.9}
$$

Note that, whereas the conditional variance of $\mathbf{X}_1$ given $\mathbf{X}_2 = x_2$ does not depend upon x_2, the conditional expectation varies linearly with x_2, with slope and intercept equal to $\frac{\sigma_1}{\sigma_2}\rho$ and $\mu_1 - \frac{\sigma_1}{\sigma_2}\rho\mu_2$, respectively.

The function described by the right-hand side of (20.8) also occurs in the context of 'linear regression' in statistics, and we take a small detour to establish the connection.

Linear Regression

20.9. Problem. Let $\mathbf{X}_1$ and $\mathbf{X}_2$ be two random variables with an unspecified joint distribution, but with finite marginal means and variances. Find the constants α and β minimizing the value of

$$E(\mathbf{X}_1 - \alpha\mathbf{X}_2 - \beta)^2. \tag{20.10}$$

The straight line with equation $y = \alpha x + \beta$, whose parameters α and β minimize (20.10) is referred to as the *regression line* in statistics.

20.10. Example. To motivate this problem, consider a situation in which for each deceased male in a certain population, a pair (x_1, x_2) of numbers has been recorded representing the following information

x_1 AGE OF THE INDIVIDUAL AT TIME OF DEATH

x_2 AGE OF HIS FATHER AT TIME OF DEATH.

A pair (x_1, x_2) may be regarded as a sample point of the joint distribution of $(\mathbf{X}_1, \mathbf{X}_2)$ in the problem described in 20.9. Let us entertain the simple minded idea of predicting the duration of the life of an individual solely on the basis of the duration of his father's life. Assume that there is an approximately linear relationship between the numbers in a pair (x_1, x_2). In other words, x_1 is approximately equal to $\alpha x_2 + \beta$, for some parameters α and β that have to be determined or estimated. A standard approach for estimating the value of such parameters relies on the so-called 'least squares' method. Suppose that the data pertaining to n individuals have been collected and let $(x_{11}, x_{21}), (x_{12}, x_{22}), \ldots, (x_{1n}, x_{2n})$ be the n recorded pairs of life spans. In the least squares method, the parameters α and β are estimated by minimizing the quantity $\sum_{i=1}^{n}(x_{1i} - \alpha x_{2i} - \beta)^2$, or equivalently, minimizing

$$\frac{1}{n} \sum_{i=1}^{n} (x_{1i} - \alpha x_{2i} - \beta)^2. \tag{20.11}$$

This equation is a statistical version of Eq. (20.10) focusing on the actual observations rather than on the underlying random variables $\mathbf{X}_1$ and $\mathbf{X}_2$.

20.11. Solution of Problem 20.9. By the hypotheses of the problem, the means and the variances of the random variables $\mathbf{X}_1$ and $\mathbf{X}_2$ exist; so let

$$\mu_1 = E(\mathbf{X}_1), \qquad\qquad \sigma_1^2 = Var(\mathbf{X}_1),$$
$$\mu_2 = E(\mathbf{X}_2), \qquad\qquad \sigma_2^2 = Var(\mathbf{X}_2).$$

Formula (20.10) can be rewritten

$$E[(\mathbf{X}_1 - \mu_1) - \alpha(\mathbf{X}_2 - \mu_2) - (\beta - \mu_1 + \alpha\mu_2)^2]. \tag{20.12}$$

(This involves adding and subtracting $\mu_1 + \alpha\mu_2$ inside the parentheses in (20.10) and gathering terms appropriately.) Developing the square in (20.12) and using the fact that the expectation is a linear operator (cf. Theorem 15.4), we obtain

$$E[(\mathbf{X}_1 - \alpha\mathbf{X}_2 - \beta)^2] =$$
$$E[(\mathbf{X}_1 - \mu_1)^2] + \alpha^2 E[(\mathbf{X}_2 - \mu_2)^2] + E\left((\beta - \mu_1 + \alpha\mu_2)^2\right)$$
$$-2\alpha E[(\mathbf{X}_1 - \mu_1)(\mathbf{X}_2 - \mu_2)] - 2E[(\mathbf{X}_1 - \mu_1)(\beta - \mu_1 + \alpha\mu_2)]$$
$$+2\alpha E[(\mathbf{X}_2 - \mu_2)(\beta - \mu_1 + \alpha\mu_2)],$$

or more compactly, since we have

$$E[(\mathbf{X}_1 - \mu_1)^2] = \sigma_1^2, \quad E[(\mathbf{X}_2 - \mu_2)^2] = \sigma_2^2,$$
$$E[(\mathbf{X}_1 - \mu_1)(\mathbf{X}_2 - \mu_2)] = Cov(\mathbf{X}_1, \mathbf{X}_2),$$
$$\text{and} \ \ E(\mathbf{X}_1 - \mu_1) = E(\mathbf{X}_2 - \mu_2) = 0,$$

$$E(\mathbf{X}_1 - \alpha\mathbf{X}_2 - \beta)^2 =$$
$$\sigma_1^2 + \alpha^2\sigma_2^2 + (\beta - \mu_1 + \alpha\mu_2)^2 - 2\alpha Cov(\mathbf{X}_1, \mathbf{X}_2). \tag{20.13}$$

Notice that β appears in the r.h.s. only in the term $(\beta - \mu_1 + \alpha\mu_2)^2$. Thus, for any value of α, $E(\mathbf{X}_1 - \alpha\mathbf{X}_2 - \beta)^2$ is smaller for $\beta = \mu_1 - \alpha\mu_2$

than for any other value of β. With $\beta = \mu_1 - \alpha\mu_2$ and $\rho = \frac{Cov(\mathbf{X}_1, \mathbf{X}_2)}{\sigma_1\sigma_2}$, Eq. (20.13) becomes

$$E(\mathbf{X}_1 - \alpha\mathbf{X}_2 - \beta)^2 = \sigma_1^2 + \alpha^2\sigma_2^2 - 2\alpha\sigma_1\sigma_2\rho$$
$$= \sigma_1^2\rho^2 - 2\alpha\sigma_1\sigma_2\rho + \alpha^2\sigma_2^2 + \sigma_1^2 - \sigma_1^2\rho^2$$
$$\text{(completing the square)}$$
$$= (\sigma_1\rho - \alpha\sigma_2)^2 + \sigma_1^2(1 - \rho^2)$$

which is minimum for α if $\sigma_1\rho = \alpha\sigma_2$. Solving the system

$$\sigma_1\rho = \alpha\sigma_2, \qquad \beta = \mu_1 - \alpha\mu_2$$

for α and β yields

$$\alpha = \frac{\sigma_1}{\sigma_2}\rho, \qquad \beta = \mu_1 - \frac{\sigma_1}{\sigma_2}\rho\mu_2.$$

This is the solution of Problem 20.9. The regression line is thus

$$y = x\rho\frac{\sigma_1}{\sigma_2} + \mu_1 - \frac{\sigma_1}{\sigma_2}\rho\mu_2.$$

A comparison with (20.8) leads us to conclude that: IF TWO RANDOM VARIABLES ARE JOINTLY NORMAL, THEN THE CONDITIONAL EXPECTATION OF $\mathbf{X}_1$ GIVEN $\mathbf{X}_2 = x_2$ COINCIDES, AS A FUNCTION OF x_2, WITH THE REGRESSION LINE.

Problems

1. Prove that the M.G.F. of a bivariate normal has the form given by Theorem 20.3.

2. Prove that the r.h.s. of (20.1) coincides with the r.h.s. of (20.4) for all real numbers x_1 and x_2 only if $\rho = 0$.

3. Fill in the details in the proof of Corollary 20.4.

4. Prove Theorem 20.5.

5. Suppose that the mean driving distance for a particular golfer is 200 yards, with a standard deviation of 15 yards. Suppose also that the mean amount of time the ball remains airborne after being hit by this golfer is 2.8 seconds, with a standard deviation of 0.3 seconds. Assume that the driving distance and the time the ball is airborne are jointly normal random variables with correlation 0.8. (a) What is the probability that this golfer hits

a drive greater than 230 yards?   (b) What is the probability that this golfer hit a drive greater than 230 yards, given that the ball remained airborne for 3.4 seconds?   (c) What is the probability that the ball remained in the air between 3.10 and 3.46 seconds, given that the ball travelled 230 yards?

6. Does the converse of Theorem 20.2 hold in general? That is, if $(\mathbf{X}_1, \mathbf{X}_2)$ is a pair of jointly distributed random variables such that the marginal densities of $\mathbf{X}_1$ and $\mathbf{X}_2$ are normal, is their joint density necessarily normal?

7. Let $\mathcal{X} = (\mathbf{X}_1, \mathbf{X}_2)$ be a pair of random variables with a normal joint density and let $M_{\mathcal{X}}$ be the joint M.G.F. Show that the covariance is given by
$$\frac{\partial^2 M_{\mathcal{X}}(0,0)}{\partial \theta_1 \partial \theta_2} - \frac{\partial M_{\mathcal{X}}(0,0)}{\partial \theta_1} \frac{\partial M_{\mathcal{X}}(0,0)}{\partial \theta_2}.$$

8. Suppose that $\mathbf{X}$ and $\mathbf{Y}$ are jointly normal random variables with respective means μ_1 and μ_2, respective variances σ_1 and σ_2, and correlation ρ; what is $\mathbb{P}(\mathbf{X} > \mathbf{Y})$?

Chapter 21

Finite Markov Chains, Basic Concepts

Loosely speaking, a Markov chain is a precise description, in the language of probability theory, of a system with a very limited memory of its past. Consider some system whose states have been observed on some not necessarily consecutive j trials numbered $n_1 < n_2 < \ldots < n_j$. (We know nothing about the states of the system on any other trials.) Suppose that the state of the system on any later trial n_{j+k} only depends upon that last observed state on trial n_j. A system satisfying such a property for any choice of trials numbers $n_1, n_2, \ldots, n_j, n_{j+k}$ is said to be 'Markovian.'[1] In words: a system is Markovian if the prediction of its future state only depends upon the last recorded state. The examples in the next two sections will pave the way to a precise definition.

A Markovian Learning Theory

21.1. Example. (TWO-STATE LEARNING.) On each trial of an experiment, a subject is presented with some material designed to induce the mastery of a given concept, fixed through the experiment. For example, an instance of the concept is presented on each trial, followed by a test. The subject's response to the test is coded as correct (**C**) or false (**F**). We assume that there only two possible states of learning: the subject can either be totally naive (state **N**),

---

or could have achieved total mastery of the concept (state **M**). Surprisingly, such a simplistic model is not so easy to reject experimentally, because the states of the subject need not be directly observable. Indeed, we suppose here that a correct response could happen in the naive state. (This can happen, for instance, if the test has a 'multiple-choice' format.) The probability of this event is denoted by γ and is referred to as the 'guessing probability.' A response arising in state **M** is always correct, however.

We assume that learning occurs according to the following rules. The subject always begins the experiment in the naive state **N**, and never leaves state **M** once this state has been achieved. The subject may grasp the concept on any trial where a correct response is given. Specifically, there is a probability θ that the subject's state changes from state **N** to state **M** in the case of a correct response. If the response is false, the subject remains in the naive state. The model has thus two parameters γ and θ which are assumed to be constant over all trials. This model is an elaboration of one encountered in 11.3. Models of this kind have been used by experimental psychologists to describe the learning behavior of humans and animals in laboratory situations (see Atkinson et al., 1965, Norman, 1972). The assumptions of the model are conveniently summarized by the *transition diagram* of Fig. 21.1, the conventions of which should be self explanatory.

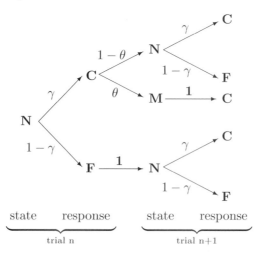

Figure 21.1: **Transition diagram in the two-state Markov learning model.**

The probabilities of the transitions between the various situations of interest are obtained by multiplying the probabilities along the corresponding paths, and summing the probabilities of the paths. For instance, the probability of occurrence of a state **N** following a state **N** is equal to

$$\gamma(1 - \theta) + (1 - \gamma) = 1 - \gamma\theta. \tag{21.1}$$

Another useful representation is provided by the directed graph of Fig. 21.2, which focuses on the subject's two possible learning states, and makes clear that the sequence of states is a Markov chain. (For graph terminology, see 23.11.) We can infer from the graph, for example, that the probability that the subject is in state **M** on trial n, given the states on some previous trials, say $n_1 < \ldots < n_j < n$, only depends upon the state on trial n_j. In fact, this probability is equal to 1 if the subject is in state **M** on trial n_j (since the subject never leaves state **M**), while this probability is equal to

$$1 - (1 - \gamma\theta)^{n - n_j} \tag{21.2}$$

if the subject is in state **N** on trial n_j, regardless of the states on trials preceding n_j.

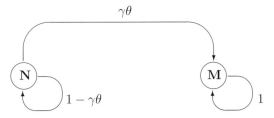

Figure 21.2: Oriented graph in the two-state Markov learning model.

Notice that the information contained in the directed graph of Fig. 21.2 may also be represented in matrix form, as in Table 21.1. Each cell of this matrix contains the probability of a transition from a state on trial n to a state on trial $n + 1$. Such a matrix is usually referred to as the 'transition matrix' of the Markov chain and is often used when dealing with finite Markov chains (that is, chains with a finite number of possible states). This representation permits a compact notation of theoretical results, and renders computation easy by the use of matrix algebra. Let us illustrate this. We denote

by P the transition matrix of Table 21.1. The transition probabilities between the states from trial n to trial $n + 2$ can then be
obtained by forming the matrix product $PP = P^2$. The resulting
matrix is given in Table 21.2.

Table 21.1: Transition matrix of the Markov chain with states N and M in Example 21.1.

Trial $n + 1$

		M	N
Trial n	M	1	0
	N	$\gamma\theta$	$1 - \gamma\theta$

Table 21.2: Matrix $PP = P^2$ of the Markov chain with states N and M in Example 21.1.

Trial $n + 2$

		M	N
Trial n	M	1	0
	N	$\gamma\theta + (1 - \gamma\theta)\gamma\theta$	$(1 - \gamma\theta)^2$

In general, the transition probabilities between the states from
trial n to trial $n + k$ are contained in the cells of the matrix P^k. In
this example, these transition probabilities only depend upon k (the
difference between n and $n + k$). We shall go back to such matters
in our discussion of the general theory.

So far, our position has been that the states of the system under investigation were identical to the two cognitive states of the
subjects, namely **M** and **N**, a viewpoint that was illustrated by the
directed graph of Fig. 21.2. But different definitions of the concept
of states are conceivable. For example, we may consider the states
of the system to be the three pairs **NF**, **NC**, and **MC** of learning
states and responses on a given trial. (The fourth pair **MF** has
probability zero and may be omitted.) We would say, for instance,
that the system is in state **NC** on trial n if the subject is naive, but
nevertheless gives a correct response on that trial.

We encounter here a terminological difficulty which is inherent to
this example: the term 'state' is used with two different meanings,

one covering the momentary mental organization of the individual in the experiment, the other describing a mathematical variable which may change its value from one trial to the next. The term 'state' is a standard one in Markov chain theory and will be used in the sequel (cf. Def. 21.5). In some situations, the states of the Markov chain under consideration will actually coincide with the mental states of the subject as postulated by the model. This is the case for the Markov chain pictured by the graph of Fig. 21.2. In general, however, the states of the subjects and of the Markov chain may be distinct, as illustrated by the Markov chain involving the three (Markov) states **NF**, **NC**, and **MC**. This Markov chain is described by the directed graph of Fig. 21.3.

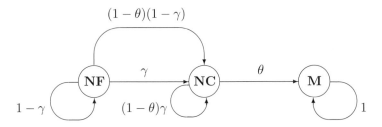

Figure 21.3: Oriented graph in the three-state Markov model.

The reader should check that the transition probabilities indicated on the graph match the basic information contained in the tree-diagram of Fig. 21.1. The corresponding transition matrix is given in Table 21.3.

Table 21.3: Transition matrix of the Markov chain with the three states NC, MC, and NF in Example 21.1.

		Trial $n+1$		
		MC	**NC**	**NF**
	MC	1	0	0
Trial n	**NC**	θ	$(1-\theta)\gamma$	$(1-\theta)(1-\gamma)$
	NF	0	γ	$1-\gamma$

More than one Markov chain may thus arise in modeling some empirical phenomenon. On the other hand, the sequence of responses of the subject (regarded as the only output of the system described by the transition diagram of Fig. 21.1), is not a Markov chain: it is easily shown that the probability of a correct response on trial n given the responses observed on some previous trials $n_1 < n_2 < \ldots < n_j$ is not equal to the probability of a correct response on trial n given the response observed on n_j (cf. Problem 1).

21.2. Remark. An important method is illustrated by the above example. When the output of some empirical system is not Markovian, it is sometimes possible to postulate unobservable states of the system, which govern the observable output, and enjoy the Markov chain property. It is as if the extensive memory of the observable past which is required to predict the next output is summarized efficiently in some unobservable state of the system. In the above example, the sequence of learning states is Markovian, while the sequence of the subject's responses is not. On occasion, this device may considerably simplify the analysis of the sytem under study (see Cox, 1955, for a discussion of this method).

21.3. Remark. In Example 21.1, the probability of a transition from state **N** to state **M** in the case of a correct response is constant over the trials. It should be clear that this is not a necessary condition for the sequence of states to be Markovian. Suppose in fact that the probability η_n of a transition from state **N** on trial n, with a correct response on that trial, to state **M** on a trial $n+1$ is nonconstant with n. Specifically, assume that

$$\eta_n = \theta \left(1 - \frac{c}{n+1} \right) \tag{21.3}$$

where $0 < \theta \leq 1$ and $0 \leq c \leq 1$ are parameters. It is easily shown that the sequence of states of the subject is a Markov chain in the sense of Definition 21.5 in the next section (see Problem 2).

As shown by our last preparatory example below, Markov chains with many possible states may be encountered.

21.4. Example. (A COIN TOSSING GAME.) Each of the two players, Peter and Paul, starts with \$100. Peter gives one dollar to Paul in the case of a head, but receives one dollar from Paul if the toss results in a tail. The game stops with the ruin of one of the two players. The total gain of Peter (which is positive or negative) is recorded on each trial and constitutes the sole output of the system. We assume that the probability of a head is some number γ, constant on every trial. A typical sequence initiating a game might be:

TRIAL NUMBER	0	1	2	3	4	5	6	7	...
RESULT OF TOSS		H	T	H	H	H	T	T	...
TOTAL GAIN OF PETER	0	-1	0	-1	-2	-3	-2	-1	...

(Notice that we begin the numbering of the trials at 0. This is a frequent practice in stochastic processes, which simplifies the writing of some results.) Let us denote by $\mathbf{G}_n$ the gain of Peter on trial n, with $n \in \mathbb{N}_0 = \{0, 1, \dots\}$. We have thus $-100 \le \mathbf{G}_n \le 100$. Notice that, with probability 1, as long as $-100 < \mathbf{G}_n < 100$, $\mathbf{G}_{n+1}$ differs from $\mathbf{G}_n$ by exactly one: we have $\mathbf{G}_{n+1} = \mathbf{G}_n + 1$ with probability $1 - \gamma$, and $\mathbf{G}_{n+1} = \mathbf{G}_n - 1$ with probability γ, regardless of the values of $\mathbf{G}_{n-1}, \mathbf{G}_{n-2}, \dots$.

The sequence $(\mathbf{G}_n)$ is a Markov chain. Note that an idealization of the data is required here. By definition, a Markov chain is an infinite sequence of random variables. Theoretically, the coin tossing game could go on forever. But it may also terminate at some finite trial. To cast this empirical situation in the framework of stochastic processes, we simply assume here that the gain of Peter remains forever at its final value (100 or -100). In fact, a similar type of idealization was required in the two-state learning model of Example 21.1 since any learning experiment would in practice stop after a finite number of trials.

The Markov chain of this example has the special property that the state on any trial is necessarily one of the two adjacent neighbors (on the lattice of integers) of the state on the preceding trial. Such a Markov chain is called a 'random walk.' The particular random walk encountered here is said to have 'two absorbing barriers,' namely, -100 and 100. Random walks are discussed in Chapter 23.

Basic Concepts

21.5. Definition. Let $\mathbf{X}_0, \mathbf{X}_1, \dots, \mathbf{X}_n, \dots$ be a sequence of random variables taking their values on the same finite set $\mathcal{S}$. Suppose that, for any positive integer n and any event E depending only on the values of the random variables $\mathbf{X}_0, \mathbf{X}_1, \dots \mathbf{X}_{n-1}$, we have, for any $i, j \in \mathcal{S}$,

$$\mathbb{P}(\mathbf{X}_{n+1} = j \,|\, \mathbf{X}_n = i, E) = \mathbb{P}(\mathbf{X}_{n+1} = j \,|\, \mathbf{X}_n = i). \qquad (21.4)$$

Then, the sequence $(\mathbf{X}_n)$ is a *(finite) Markov chain*. Note that in this case we also have

$$\mathbb{P}(\mathbf{X}_{n+1} = j \,|\, \mathbf{X}_n = i, \dots , \mathbf{X}_1, \mathbf{X}_0) = \mathbb{P}(\mathbf{X}_{n+1} = j \,|\, \mathbf{X}_n = i). \quad (21.5)$$

The formulations of the Markovian property in terms of (21.4) and (21.5) are actually equivalent (see Problem 5.) The set $\mathcal{S}$ is the *state space* of the chain, and the elements of $\mathcal{S}$ are called the *states*. We shall use the abbreviations

$$p_n(i) = \mathbb{P}(\mathbf{X}_n = i) \qquad\qquad\qquad (21.6)$$
$$t_{n,m}(i, j) = \mathbb{P}(\mathbf{X}_m = j \,|\, \mathbf{X}_n = i). \qquad\qquad (21.7)$$

Thus, Eq. (21.4) can be rewritten as

$$\mathbb{P}(\mathbf{X}_{n+1} = j \,|\, \mathbf{X}_n = i, E) = t_{n,n+1}(i, j).$$

The quantities $t_{n,m}(i, j)$ are referred to as the *transition probabilities* of the chain $(\mathbf{X}_n)$. Defining the chain requires thus specifying the *initial probabilities* $p_0(i)$ of the states and the transition probabilities $t_{n,n+1}(i, j)$.

21.6. Remark. The simplicity of the definition of a Markov chain is deceptive. A casual approach may lead to an intuitive coding of (21.4) into the statement 'the events on trial $n+1$ only depend upon the events on trial n.' In turn, this might lead to infer, for example, that if $(\mathbf{G}_n)$ is a Markov chain, then for any state i, j, and k, and with the event E as in Eq. (21.4),

$$\mathbb{P}(\mathbf{G}_{n+1} = j \,|\, \mathbf{G}_n = i \text{ or } \mathbf{G}_n = k, E)$$
$$= \mathbb{P}(\mathbf{G}_{n+1} = j \,|\, \mathbf{G}_n = i \text{ or } \mathbf{G}_n = k).$$

In fact, this equation is not derivable from (21.4). Here is a counterexample. Let $\mathbf{G}_n$ denote the random variables of Example 21.4. Consider the case

$$\mathbb{P}(\mathbf{G}_3 = 1 \,|\, \mathbf{G}_2 = 2 \text{ or } \mathbf{G}_2 = 0, E)$$
$$= \mathbb{P}(\mathbf{G}_3 = 1 \,|\, \mathbf{G}_2 = 2 \text{ or } \mathbf{G}_2 = 0), \qquad (21.8)$$

and suppose that E denotes the event '$\mathbf{G}_1 = -1$.' It is easy to verify that the two sides of the above equation give different results (see Problem 9).

The following result is fundamental.

21.7. Theorem. *For any trial numbers $0 \leq n < m < r$ and any states i and j in $\mathcal{S}$,*

$$t_{n,r}(i,j) = \sum_{k \in \mathcal{S}} t_{n,m}(i,k) t_{m,r}(k,j). \tag{21.9}$$

Equation (21.9) is known as the *Chapman-Kolmogorov Equation.*

PROOF. This result follows immediately from the Theorem of Total Probabilities (see 9.8) and the definition of a Markov chain. We have by definition

$$\begin{aligned} t_{n,r}(i,j) &= \mathbb{P}(\mathbf{X}_r = j \mid \mathbf{X}_n = i) \\ &= \sum_{k \in \mathcal{S}} \mathbb{P}(\mathbf{X}_r = j \mid \mathbf{X}_m = k, \mathbf{X}_n = i) \mathbb{P}(\mathbf{X}_m = k \mid \mathbf{X}_n = i) \\ &= \sum_{k \in \mathcal{S}} \mathbb{P}(\mathbf{X}_r = j \mid \mathbf{X}_m = k,) \mathbb{P}(\mathbf{X}_m = k \mid \mathbf{X}_n = i) \\ &= \sum_{k \in \mathcal{S}} t_{n,m}(i,k) t_{m,r}(k,j). \end{aligned}$$

$\square$

We mention an immediate but useful consequence.

21.8. Corollary. *For any trial numbers $0 \leq n < m < r$, and any states i, j, and k in $\mathcal{S}$,*

$$t_{n,r}(i,j) \geq t_{n,m}(i,k) t_{m,r}(k,j). \tag{21.10}$$

21.9. Matrix Notation. Results concerning finite Markov chains are conveniently written in the notation of vectors and matrices. Let the elements of a state space $\mathcal{S}$ be numbered $1, 2, \dots, q$. Thus, the variables i, k, and j in Eqs. (21.9) and (21.10) run in the set $\{1, 2, \dots, q\}$. Let

$$T_{n,m} = \begin{pmatrix} t_{n,m}(1,1) & t_{n,m}(1,2) & \cdots & t_{n,m}(1,q) \\ t_{n,m}(2,1) & t_{n,m}(2,2) & \cdots & t_{n,m}(2,q) \\ \cdots & \cdots & \cdots & \cdots \\ \cdots & \cdots & \cdots & \cdots \\ t_{n,m}(q,1) & t_{n,m}(q,2) & \cdots & t_{n,m}(q,q) \end{pmatrix} \tag{21.11}$$

denote a square matrix, the cell (i, j) of which contains the probability of a transition from state i on trial n to state j on trial m. Note that the probabilities of each row of the matrix $T_{n,m}$ sum to one. Such a matrix is often called a *transition matrix*, or a *stochastic matrix*, examples of which were given in Tables 21.1 and 21.3. Extending the notation introduced in (21.6), we also write

$$\mathbf{p}_n = \big(p_n(1), \ldots, p_n(t)\big) \tag{21.12}$$

for the vector of the state probabilities on trial n. In the notation of (21.11) and (21.12), the Chapman-Kolmogorov Equation (cf. Theorem 21.7) has the compact expression

$$T_{n,r} = T_{n,m} T_{m,r}, \tag{21.13}$$

for any positive integers $n < m < r$. A simple expression is also available for the vector of the state probabilities on trial $n + 1$ as a function of the state probabilities on trial n and of the transition probabilities. We clearly have, for any trial number n and any state i in $\mathcal{S}$,

$$\mathbf{p}_{n+1}(i) = \sum_{k \in \mathcal{S}} p_n(k) t_{n,n+1}(k, i).$$

(This is an application of the Theorem of Total Probabilities; cf. Theorem 9.8.) Using the standard notation for the product of a vector by a matrix, this becomes

$$\mathbf{p}_{n+1} = \mathbf{p}_n T_{n,n+1}.$$

By induction, it follows that

$$\mathbf{p}_{n+m} = \mathbf{p}_m T_{m,m+1} T_{m+1,m+2} \cdots T_{n-1,n}.$$

Thus, in particular,

$$\mathbf{p}_n = \mathbf{p}_0 T_{0,1} T_{1,2} \cdots T_{n-1,n}. \tag{21.14}$$

Of particular interest in the sequel is the situation where all the one-step transition matrices $T_{n,n+1}$ are identical to some basic transition matrix $\mathcal{T}$. Equation (21.14) takes then the very simple form

$$p_n = p_0 \mathcal{T}^n,$$

where $\mathcal{T}^n$ stands for the nth power of the matrix $\mathcal{T}$. This means that the probabilities of the states on each trial can be computed from the initial vector p_0 and the successive powers of the matrix $\mathcal{T}$. An example of such a situation was encountered in Example 21.1. Chapter 22 is devoted to this special case.

Problems

1. In the system described by the tree-diagram of Fig. 21.1 (Example 21.1), prove that the sequence of observable responses of the subject (correct-false) is not a Markov chain.

2. In the model specified by Eq. (21.3) in Remark 21.3, the probability of a transition from state **N** to state **M** in the case of a correct response is nonconstant over the trials. Show that the sequence of learning states is nevertheless a Markov chain in the sense of Definition 21.5. You should begin by casting the model in terms of the notation of Definition 21.5. (Start with defining the appropriate sequences of random variables; for example, set $\mathbf{X}_n = 1$, if the response on trial n is correct, and $\mathbf{X}_n = 0$ otherwise, etc.)

3. (Continuation.) Compute the conditional probabilities that the subject is in learning state **M** on trial n given the states on previous trials $n_1 < n_2 < \ldots < n_j < n$. Write the transition matrix of this process. Compute also the probability of a correct response on trial n.

4. Compute the correlation between $\mathbf{X}_n$ and $\mathbf{X}_{n+1}$ (cf. 16.5).

5. Prove that the two formulations of the Markovian property for a sequence of random variables embodied in Eqs. (21.4) and (21.5) are equivalent. (Hint: Begin by examining a special case with five random variables $\mathbf{X}_1, \ldots, \mathbf{X}_5$. Based on this example, set up a suitable formalism for the general situation.)

6. Suppose that, in the setting of Example 21.1, the subject is in state **N** on trial 7. What is the probability that she is in state **M** on trial 10? That she is in state **N** on trial 10?

7. A gambler playing roulette bets on one of two numbers on each play, each number paying off with probability p. If he wins on a play, he bets on the same number the next play, otherwise, he bets on the other number. Let the two states be the two different bets that he makes.
   (a) Find the transition matrix $\mathcal{T}$.
   (b) Show that

$$\mathcal{T}^n = \begin{pmatrix} \frac{1}{2} + \frac{1}{2}(2p-1)^n & \frac{1}{2} - \frac{1}{2}(2p-1)^n \\ \frac{1}{2} - \frac{1}{2}(2p-1)^n & \frac{1}{2} + \frac{1}{2}(2p-1)^n \end{pmatrix}.$$

   (c) What can you say about $\lim_{n\to\infty} \mathcal{T}^n$? What does this mean in terms of the states?

8. Archie and Nero sit facing each other, each holding two playing cards. There are two red cards and two black cards. Archie draws a card, at random, from Nero's hand and puts it in his own, and Nero simultaneously does the same, drawing a card from Archie's hand and putting it in his own. The two repeat this card switching indefinitely, noting after each switch the number of red cards in Nero's hand. (Thus, the states are the number of red cards in Nero's hand.) (a) Find the transition matrix. (b) What is the probability that Nero has one red card in his hand after the second switch, given that he begins with one red card in his hand? (c) Suppose that the probability that Nero begins with one red card in his hand is $\frac{1}{8}$ and the probability that he begins with two red cards in his hand is $\frac{1}{2}$. Find the probability that he has no red cards in his hand after the second switch.

9. Justify the contention in Remark 21.6 by checking that the two sides of Equation (21.8) indeed give different results.

Chapter 22

Homogeneous Markov Chains

Notation and Basic Result

22.1. Definition. In Remark 21.3, we discussed a case where the one-step transition probabilities $t_{n,n+1}(i,j)$ in a finite Markov chain $(\mathbf{X}_n)$ were dependent upon the trial number n. When, for all pairs of states (i,j), these transitions do not vary with the trial number, the Markov chain $(\mathbf{X}_n)$ is called *homogeneous* or, equivalently, is said to have *stationary transition probabilities*. In this case, there are numbers $\tau(i,j)$ specified for every i, j in the state space $\mathcal{S}$ of the Markov chain, such that, for all trial numbers $n \geq 0$,

$$\tau(i,j) = \mathbb{P}(\mathbf{X}_{n+1} = j \mid \mathbf{X}_n = i).$$

We have thus

1. $\tau(i,j) \geq 0$, for all $i, j \in S$;

2. $\sum_{j \in S} \tau(i,j) = 1$, for all $i \in S$.

In other terms, there is a matrix $\mathcal{T} = \big(\tau(i,j)\big)$ such that, for all integers $n \geq 0$ and with $T_{n,m}$ as in 21.9,

$$\mathcal{T} = T_{n,n+1}. \tag{22.1}$$

The matrix $\mathcal{T}$ is called the *(one-step) transition matrix* of the chain.

22.2. Theorem. *In an homogeneous Markov chain $(\mathbf{X}_n)$ with a transition matrix $\mathcal{T}$, the transition probabilities $t_{n,m}(i,j)$ only depend upon the difference $m - n$. Specifically, we have*

$$T_{n,m} = \mathcal{T}^{m-n}. \tag{22.2}$$

Moreover, for any trial number $n \geq 1$,

$$\mathbf{p}_n = \mathbf{p}_0 \mathcal{T}^n \tag{22.3}$$

where $\mathbf{p}_0$ is the initial vector of probabilities of the states, and $\mathbf{p}_n$ is the vector of probabilities of those states on trial n.

In view of this result, it makes sense to write $\tau_k(i,j)$ for the conditional probability of a transition from state i on trial n to state j on trial $n + k$, for any nonnegative integers n; formally,

$$\tau_k(i,j) = \mathbb{P}(\mathbf{X}_{n+k} = j \mid \mathbf{X}_n = i). \tag{22.4}$$

Note that $\tau_k(i,j)$ is the entry in the cell (i,j) of the matrix $\mathcal{T}^k$ and satisfies $\tau_k(i,j) = t_{n,n+k}(i,j)$ for every integer $n \geq 0$ (cf. Eq. (21.7)).

PROOF OF THEOREM 22.2. Equation (22.2) follows from Equation (21.13), which yields

$$T_{n,m} = T_{n,n+1} T_{n+1,n+2} \cdots T_{m-1,m} = \mathcal{T}^{m-n},$$

since, by Eq. (22.1), each of the factors in the product is equal to $\mathcal{T}$.

Equation (22.3) is then obtained from the fact that, by definition of $T_{0,n}$ and by Eq. (22.2), we get $\mathbf{p}_n = \mathbf{p}_0 T_{0,n} = \mathbf{p}_0 \mathcal{T}^n$. $\qquad \square$

Some typical questions arise when investigating Markov chains; for example: Which states of the state space can lead to some other state? Are there 'disappearing' states, that is, states that are no longer 'visited' after some time? Are there sets of states such that, whenever the chain 'visits' the set, it is trapped and cannot get out? Are there states, or sets of states, to which the chain will necessarily go? A discussion of such matters requires a precise terminology, a key concept of which is a particular ordering of the states, which is defined in the next section.

22.3. Convention. Except when explicitly indicated, we consider in the rest of this chapter a finite homogeneous Markov chain $(\mathbf{X}_n)$ with state space $\mathcal{S} = \{1, 2, \ldots, q\}$, initial vector $\mathbf{p}_0$, transition matrix $\mathcal{T} = (\tau(i,j))$, and n-step transition matrix $\mathcal{T}^n = (\tau_n(i,j))$. We recall that the trials are numbered from zero.

Ordering the States

22.4. Definition. Let $\rightarrowtail$ be a binary relation (cf. 1.7) on the set of states $\mathcal{S}$, defined by

$$i \rightarrowtail j \quad \Longleftrightarrow \quad i = j \text{ or } \tau_n(i,j) > 0 \text{ for some integer } n > 0.$$

When $i \rightarrowtail j$, we shall sometimes say that j is *accessible* from i, or i *communicates* with j. This relation will be used to gather the states into classes of mutually accessible states. The relation $\rightarrowtail$ is the *accessibility* relation of the Markov chain $(\mathbf{X}_n)$. A state j such that $j \rightarrowtail k \rightarrowtail j$ for some state k is called a *return* state.

22.5. Theorem. *The relation $\rightarrowtail$ is a quasiorder on the set $\mathcal{S}$ of states, that is, $\rightarrowtail$ is reflexive and transitive on $\mathcal{S}$.*

(See Roberts, 1979, or Trotter, 1992, for the terminology of order relations.) The reflexivity of $\rightarrowtail$ is immediate from its definition, and the transitivity results from Corollary 21.8.

22.6. Example. Consider the Markov chain represented by the directed graph of Fig. 22.1, with state space $\{1,2,3,4,5,6\}$. All the states are accessible from state 1. (We have $1 \rightarrowtail i$ for $i = 1, \dots, 6$.)

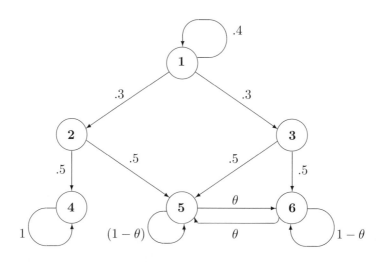

Figure 22.1: Oriented graph of the homogeneous Markov chain of Example 22.6. We assume that $0 < \theta \le 1$.

Since only states 5 and 6 are mutually accessible, we have five classes of states: $\{1\}$, $\{2\}$, $\{3\}$, $\{4\}$, and $\{5,6\}$. There are two 'terminal' classes of states, namely, $\{4\}$ and $\{5,6\}$. No matter what the initial vector $\mathbf{p}_0$ is, the chain will ultimately, with probability one, get trapped into either class $\{4\}$ or class $\{5,6\}$. Let us formalize some of these ideas.

For Markov chain terminology in the next section, we follow Kemeny and Snell (1960).

Ergodic and Transient States

22.7. Definition. Let $\mathbb{S}$ be the set of equivalence classes induced by the accessibility relation $\rightarrowtail$, and let $\precsim$ be the corresponding partial order on the set $\mathbb{S}$ (cf. 1.7). For any state i, we denote by $[i]$ the class of all states j such that both $i \rightarrowtail j$ and $j \rightarrowtail i$ hold. Thus,

$$\mathbb{S} = \{[i] \mid i \in \mathcal{S}\},$$

and for all $i, j \in \mathcal{S}$,

$$[i] \precsim [j] \quad \Longleftrightarrow \quad i \rightarrowtail j.$$

The elements of $\mathbb{S}$ are thus classes of states. Every maximal element of $\mathbb{S}$ (for the partial order $\precsim$) is an *ergodic class*, and the elements of an ergodic class are called *ergodic states*. An element of $\mathbb{S}$ which is not ergodic is said to be a *transient class*. An element of a transient class is called a *transient* state. We denote by $\mathcal{E}$ and $\mathbb{E}$, respectively, the set of all ergodic states and the family of all ergodic classes. The letters $\mathcal{U}$ and $\mathbb{U}$ stand for the set of all transient states and the family of all transient classes, respectively. Thus,

$$\mathcal{S} = \mathcal{E} + \mathcal{U} \quad \text{and} \quad \mathbb{S} = \mathbb{E} + \mathbb{U}.$$

When an ergodic class contains a single state, this state is called *absorbing*.

In Example 22.6, there are two ergodic classes, $\{4\}$ and $\{5,6\}$, and three transient classes $\{1\}$, $\{2\}$, and $\{3\}$. The states 4, 5, and 6 are ergodic, while the thee other states are transient. We have $\mathbb{E} = \{\{4\},\{5,6\}\}$ and $\mathbb{U} = \{\{1\},\{2\},\{3\}\}$. State 4 is the only absorbing state. Since any partial order on a finite set has at least one maximal element, we obtain the following theorem:

22.8. Theorem. *Any finite Markov chain has at least one ergodic state. However, there may not be any transient states.*

We also have:

22.9. Theorem. *A state i is absorbing if and only if $\tau(i, i) = 1$.*

PROOF. If the state i is absorbing, then $[i] = \{i\}$ and is maximal for the partial order $\precsim$. Thus, $\tau(i, j) = 0$ for all states $j \neq i$. Since $\sum_{k \in \mathcal{S}} \tau(i, k) = 1$, we have $\tau(i, i) = 1$.

Conversely, $\tau(i, i) = 1$ implies that $\tau(i, j) = 0$ for all states $j \neq i$. This implies that $[i] = \{i\}$ and is maximal. Thus, i is an absorbing state. $\qquad\qquad\qquad\qquad\qquad\qquad\qquad\qquad\qquad\qquad\square$

In general, a Markov chain will tend to evolve towards the ergodic states. Formally:

22.10. Theorem. *For the finite Markov chain $(\mathbf{X}_n)$ with the family $\mathbb{E}$ of ergodic classes, we have*

$$\lim_{n \to \infty} \mathbb{P}(\mathbf{X}_n \in \cup \mathbb{E}) = 1.$$

PROOF. Since the events $\cup \mathbb{E}$ and $\cup \mathbb{U}$ are disjoint and their union is the sample space, this theorem holds if

$$\lim_{n \to \infty} \mathbb{P}(\mathbf{X}_n \in \cup \mathbb{U}) = 0.$$

For any trial n, the event $\mathbf{X}_n \in \cup \mathbb{U}$ is a subset of the event $\mathbf{X}_0 \in \cup \mathbb{U}$, which yields

$$\begin{aligned}
\mathbb{P}(\mathbf{X}_n \in \cup \mathbb{U}) &= \mathbb{P}\big((\mathbf{X}_n \in \cup \mathbb{U}) \cap (\mathbf{X}_0 \in \cup \mathbb{U})\big) \\
&= \mathbb{P}(\mathbf{X}_n \in \cup \mathbb{U} \mid \mathbf{X}_0 \in \cup \mathbb{U}) \mathbb{P}(\mathbf{X}_0 \in \cup \mathbb{U}).
\end{aligned}$$

It suffices thus to show that the sequence

$$a_n = \mathbb{P}(\mathbf{X}_n \in \cup \mathbb{U} \mid \mathbf{X}_0 \in \cup \mathbb{U})$$

converges to zero. Notice that (a_n) converges. Indeed, it is bounded from below and not increasing because

$$\begin{aligned}
a_{n+1} &= \mathbb{P}(\mathbf{X}_{n+1} \in \cup \mathbb{U} \mid \mathbf{X}_0 \in \cup \mathbb{U}) \\
&= \mathbb{P}(\mathbf{X}_{n+1} \in \cup \mathbb{U} \mid \mathbf{X}_n \in \cup \mathbb{U}, \mathbf{X}_0 \in \cup \mathbb{U}) \mathbb{P}(\mathbf{X}_n \in \cup \mathbb{U} \mid \mathbf{X}_0 \in \cup \mathbb{U}) \\
&= \mathbb{P}(\mathbf{X}_{n+1} \in \cup \mathbb{U} \mid \mathbf{X}_n \in \cup \mathbb{U}) \, a_n \\
&\leq a_n.
\end{aligned}$$

We now show that (a_n) has a subsequence converging to zero. Suppose that $\mathbf{X}_0$ is in some transient class. Then, this class is not a maximal element of the partial order $\precsim$, and there exists at least one finite sequence of states $i_0 \longmapsto i_1 \longmapsto \dots \longmapsto i_\ell$ leading from the transient state i_0 visited by $\mathbf{X}_0$ to some ergodic state i_ℓ. For any transient state i, let m_i be the length of the shortest sequence leading from i to any ergodic state, and let q_i be the probability of reaching some ergodic state in m_i steps. Let

$$m = \max\{m_i \mid i \in \cup\mathbb{U}\}, \qquad q = \min\{q_i \mid i \in \cup\mathbb{U}\}.$$

Then

$$\mathbb{P}(\mathbf{X}_{m_i} \in \cup\mathbb{U} \mid \mathbf{X}_0 = i) = (1 - q_i) \leq (1 - q),$$

yielding

$$a_m = \mathbb{P}(\mathbf{X}_m \in \cup\mathbb{U} \mid \mathbf{X}_0 = i) \leq (1 - q) < 1,$$

and so, for any positive integer k,

$$a_{mk} = \mathbb{P}(\mathbf{X}_{mk} \in \cup\mathbb{U} \mid \mathbf{X}_0 = i) \leq (1 - q)^k < 1.$$

Thus, $\lim_{n \to \infty} a_n = 0$, and the results follows. $\qquad\qquad\square$

First Visits and Occupation Times

In many empirical situations, only one realization of the process assumed to be a Markov chain can be observed. In such a case, computing statistics concerning the visits of the various states in the course of this realization can be revealing of the structure of the chain. A number of relevant tools are introduced in this section.

22.11. Definition. We begin by defining, for every state i of a Markov chain $(\mathbf{X}_n)$, a sequence of random variables

$$\mathbf{V}_{n,i} = \begin{cases} 1 & \text{if } \mathbf{X}_n = i, \\ 0 & \text{if } \mathbf{X}_n \neq i, \end{cases}$$

for $n = 0, 1, \dots$. We also define, for $n = 0, 1, \dots$ and any state $i \in \mathcal{S}$, the 'first visit' event $\mathbf{F}_{n,i}$, which occurs whenever $\mathbf{V}_{0,i} = \mathbf{V}_{1,i} = \dots = \mathbf{V}_{n-1,i} = 0$ and $\mathbf{V}_{n,i} = 1$. (Thus, $\mathbf{F}_{n,i}$ is realized whenever n is the first trial on which the chain visits state i; in particular, $\mathbf{F}_{0,i}$ is realized when $\mathbf{X}_0 = i$.) Notice that, for a fixed i, all the events $\mathbf{F}_{n,i}$ are disjoint and so $\mathbb{P}(\cup_{n=0}^{\infty}\mathbf{F}_{n,i}) = \sum_{n=0}^{\infty}\mathbb{P}(\mathbf{F}_{n,i}) \leq 1$. Finally, we also define for $n > 0$ the conditional probability

$$f_n(i,j) = \mathbb{P}(\mathbf{F}_{n,j} \mid \mathbf{X}_0 = i) \tag{22.5}$$

that the first visit of state j, given that the chain was initially in state i, occurs in exactly n steps, together with the limit

$$f(i,j) = \sum_{n=1}^{\infty} f_n(i,j).$$

Intuitively, $f(i,j)$ is the probability that the chain will ever visit state j if it starts in state i. There are numerous relations between these various concepts, and some of them are discussed in the rest of this section. We begin with a simple result linking first visits and probabilities.

22.12. Theorem. *Writing by convention* $\tau_0(j,j) = 1$, *we have for any states* i *and* j *and any integer* $n > 1$

$$\tau_n(i,j) = \sum_{k=1}^{n} f_k(i,j)\tau_{n-k}(j,j). \tag{22.6}$$

Equation (22.6) is sometimes referred to as the *first visit* or *first entrance* decomposition formula.

PROOF. The event $\mathbf{X}_n = j$ is the disjoint union of all the events $(\mathbf{X}_n = j, \mathbf{F}_{k,j})$, for $0 \le k \le \infty$. Notice, however, that all the events $(\mathbf{X}_n = j, \mathbf{F}_{k,j})$ for $k > n$ have measure zero. Using (22.5) and (22.4), we get successively

$$\tau_n(i,j) = \mathbb{P}(\mathbf{X}_n = j \mid \mathbf{X}_0 = i)$$

$$= \sum_{k=0}^{\infty} \mathbb{P}(\mathbf{X}_n = j, \mathbf{F}_{k,j} \mid \mathbf{X}_0 = i)$$

$$= \sum_{k=0}^{n} \mathbb{P}(\mathbf{X}_n = j, \mathbf{F}_{k,j} \mid \mathbf{X}_0 = i)$$

$$= \sum_{k=0}^{n} \mathbb{P}(\mathbf{F}_{k,j} \mid \mathbf{X}_0 = i)\mathbb{P}(\mathbf{X}_n = j \mid \mathbf{X}_k = i)$$

$$= \sum_{k=0}^{n} f_k(i,j)\tau_{n-k}(j,j). \qquad \square$$

We consider next the connection between the ergodicity of a state j and the behavior of the series $\sum_{n=1}^{\infty} \tau_n(j,j)$. Assume that j is an ergodic state. Thus, j belongs to the ergodic set $[j]$. Loosely speaking, once entering the ergodic set $[j]$, the chain is trapped in it, and only those states belonging to $[j]$ will be accessed. Because $[j]$ is finite, this means that the chain will always eventually return to state j. In other words $\tau_n(j, j)$ will not vanish. If the series is bounded above, it must converge. But then (by a standard result of calculus) we must have $\lim_{n\to\infty} \tau_n(j, j) = 0$, and we obtain a contradiction. This heuristic argument suggests that, for any ergodic state j, we must have

$$\sum_{n=1}^{\infty} \tau_n(j, j) = \infty. \tag{22.7}$$

Actually, Condition (22.7) is equivalent to the ergodicity of j.

22.13. Definition. A state j is *persistent* if (22.7) holds. We use the abbreviation $f(j) = f(j, j)$ and we say that j is *recurrent* if $f(j) = 1$. Otherwise, state j is called *nonrecurrent*.

The theorem below summarizes and completes our discussion. No proof is given here.

22.14. Theorem. *The following three conditions are equivalent for any state j:*

(i) *state j is ergodic;*

(ii) *state j is persistent;*

(iii) *state j is recurrent.*

Moreover, if $f(j) < 1$, then

$$\sum_{n=1}^{\infty} \tau_n(j, j) = \frac{1}{1 - f(j)}.$$

Classification of Homogeneous Markov Chains

22.15. Definition. We recall that a return state is a state j such that $j \rightarrowtail i \rightarrowtail j$ for some state $i \neq j$. Thus, j is a return state if the set

$$R(j) = \{n \in \mathbb{N} \mid \tau_n(j, j) > 0\}$$

is not empty. In such a case, the set $R(j)$ is necessarily infinite since we obviously have, for any positive integers n and m

$$n, m \in R(j) \qquad \text{implies} \qquad (n + m) \in R(j),$$

that is, the set of positive integers $R(j)$ is closed under addition. The *period* of a return state j is the greatest common divisor of $R(j)$. A state is said to be *aperiodic* if it has period 1. A homogeneous Markov chain is *aperiodic* if all its states are aperiodic. Note that if two states are mutually accessible, then they necessarily have the same period (see Problem 3).

A chain without transient classes is called *ergodic*. Such a chain may have several ergodic classes, however. When a chain has a unique ergodic class, then this chain is said to be *irreducible*. The justification for this terminology is that a chain with several ergodic classes may be decomposed into components chains, each with a single ergodic class. Each of these subchains may be studied separately since there is no communication between the ergodic classes.

We follow Kemeny and Snell (1960) in calling *regular* a homogeneous, finite Markov chain which is ergodic, irreducible, and aperiodic. The next section is devoted to a discussion of such chains.

Regular Markov Chains

22.16. Theorem. *A Markov chain* $(\mathbf{X}_n)$ *is regular if and only if there is a positive integer N such that whenever $n \geq N$, then $\tau_n(i,j) > 0$ for all states i, j in $\mathcal{S}$.*

Thus, in the conditions of the theorem, all entries of the matrix $\mathcal{T}^n$ are positive for any $n \geq N$.

PROOF. Suppose that $(\mathbf{X}_n)$ is regular. Then, it is aperiodic and all the states have period one. Since the set of states is finite, there exists necessarily a positive integer K such that $\tau_n(i,i) > 0$ for all states i and all $n \geq K$. The chain being also irreducible, all states are mutually accessible. This means that there is for every pair of states (i,j) a positive integer m_{ij} satisfying $\tau_{m_{ij}}(i,j) > 0$. Using Corollary 21.8, we obtain for $n \geq K$

$$\tau_{n+m_{ij}}(i,j) \geq \tau_n(i,i)\tau_{m_{ij}}(i,j) > 0.$$

Let m be any common multiple of the numbers m_{ij} (for all $i, j \in S$). By induction, we obtain $\tau_{n+m}(i,j) > 0$ for all $n \geq K$ and all states i and j. Setting $N = m + K$, all the entries of the matrix T^N are positive and so are the entries of any matrix T^n for $n > N$, as is easily verified. The converse follows immediately from the definitions. $\square$

The next theorem concerns the convergence of the transition probabilities for large n and is the fundamental result for regular Markov chains.

22.17. Theorem. *For any state j in a regular Markov chain, there is a number $\alpha_j > 0$ such that $\sum_{j \in S} \alpha_j = 1$, and for any pair (i, j) of states, we have*

$$\lim_{n \to \infty} \tau_n(i, j) = \alpha_j.$$

In other terms, the powers T^n of the transition matrix T are converging to a stochastic matrix A, each of the $q = |S|$ rows of which is the same vector $\boldsymbol{\alpha} = (\alpha_1, \dots, \alpha_q)$.

To establish this result, we show that for $1 \leq j \leq q$, the jth column vector of the matrix T^n tends to a vector containing all identical terms α_j; that is, as $n \to \infty$, we have

$$\begin{pmatrix} \tau_n(1, j) \\ \tau_n(2, j) \\ \dots \\ \dots \\ \tau_n(q, j) \end{pmatrix} \quad \rightarrow \quad \begin{pmatrix} \alpha_j \\ \alpha_j \\ \dots \\ \dots \\ \alpha_j \end{pmatrix}.$$

The main argument is based on the following preparatory result.

22.18. Lemma. *Let $T = (\tau(i, j))$, $1 \leq i \leq q$, $1 \leq j \leq q$, be a stochastic matrix with positive entries. For any index j, let $M_n(j)$ be the maximum value in the jth column vector of T^n, and let $m_n(j)$ be the minimum value in that column vector. Then, the sequence $m_n(j)$ is nondecreasing and the sequence $M_n(j)$ is nonincreasing. Accordingly, the sequence*

$$r_n(j) = M_n(j) - m_n(j)$$

specifying the range of values in that column vector is nonincreasing and in fact tends to zero.

PROOF. Let (i, j) be any pair of states, and let δ be the smallest entry in the matrix $\mathcal{T}$. Witout loss of generality, suppose that $m_n(j) = \tau_n(1, j)$. Using the Chapman-Kolmogorov Equation (Theorem 21.7), we obtain successively

$$
\begin{aligned}
\tau_{n+1}(i, j) &= \sum_{k \in \mathcal{S}} \tau_1(i, k) \tau_n(k, j) \\
&\leq \tau_1(i, 1) m_n(j) + (1 - \tau_1(i, 1)) M_n(j) \\
&\leq \delta m_n(j) + (1 - \delta) M_n(j) \\
&= M_n(j) - \delta (M_n(j) - m_n(j)) .
\end{aligned}
$$

We have in particular

$$
M_{n+1}(j) \leq M_n(j) - \delta (M_n(j) - m_n(j)) \tag{22.8}
$$

which shows that the sequence $M_n(j)$ is nonincreasing. A similar argument applied to the sequence $m_n(j)$ yields

$$
m_{n+1} \geq m_n(j) + \delta (M_n(j) - m_n(j)) ,
$$

or, equivalently,

$$
-m_{n+1} \leq -m_n(j) - \delta (M_n(j) - m_n(j)) . \tag{22.9}
$$

Adding (22.8) and (22.9) and grouping terms, we obtain

$$
\begin{aligned}
r_{n+1}(j) &= M_{n+1}(j) - m_{n+1}(j) \\
&\leq M_n(j) - m_n(j) - 2\delta (M_n(j) - m_n(j)) \\
&= (M_n(j) - m_n(j)) (1 - 2\delta) \\
&= r_n(j)(1 - 2\delta) .
\end{aligned}
$$

Hence for $n > 1$,

$$
r_n(j) \leq r_1(j)(1 - 2\delta)^{n-1} . \tag{22.10}
$$

Since $0 < 2\delta < 1$, we have $r_n(j) \to 0$, as asserted. $\square$

PROOF OF THEOREM 22.17. If the entries of the transition matrix of the Markov chain $(\mathbf{X}_n)$ are all positive, the result follows from Lemma 22.18. Indeed, the two sequences $m_n(j)$ and $M_n(j)$ converge, since they are monotonic and bounded. Because the range $r_n(j)$ tends to zero, these two sequences must converge to the same limit, which we denote by α_j. By definition of $m_n(j)$ and $M_n(j)$, we have $0 < m_n(j) \le \tau_n(i, j) \le M_n(j)$ for any pair of states (i, j), yielding $\lim_{n \to \infty} \tau_n(i, j) = \beta_j$.

In general, the matrix $\mathcal{T}$ may have some zero entries. However, by Theorem 22.16, there is a positive integer N such that all the entries of $\mathcal{T}^N$ are positive. Applying Lemma 22.18 to the matrix $\mathcal{T}^N$, it follows that, for any state j, the sequence $r_{Nn}(j)$ tends to zero as $n \to \infty$. Because $r_n(j)$ is nonincreasing, we must conclude that it tends to zero. The rest of the argument is as in the first paragraph of this proof.   □

The next result completes the picture.

22.19. Theorem. Let $\mathbf{p}_0 = (p_0(1), \dots, p_0(q))$ be the initial vector of state probabilities, and suppose that all the conditions of Theorem 22.17 are statisfied, with in particular $\mathcal{T}^n \to \mathcal{A}$. Then:

(i) $\mathbf{p}_0 \mathcal{T}^n$ tends to the vector α, for any arbitrarily chosen vector $\mathbf{p}_0$;

(ii) α is the unique vector satisfying the equation $\alpha \mathcal{T} = \alpha$;

(iii) $\mathcal{A} \mathcal{T} = \mathcal{T} \mathcal{A} = \mathcal{A}$.

PROOF. Because all the rows of the matrix $\mathcal{A}$ are identical to α and $\sum_{j \in S} p_0(j) = 1$, we have $\mathbf{p}_0 \mathcal{A} = \alpha$. This leads to

$$\lim_{n \to \infty} \mathbf{p}_0 \mathcal{T}^n = \mathbf{p}_0 \mathcal{A} = \alpha,$$

which proves (i). Let $\mathbf{q}$ be any vector satisfying $\mathbf{q}\mathcal{T} = \mathbf{q}$. By induction, we get $\mathbf{q}\mathcal{T}^n = \mathbf{q}$. Using (i), we obtain $\lim_{n \to \infty} \mathbf{q}\mathcal{T}^n = \alpha = \mathbf{q}$, yielding (ii).

Finally, (iii) follows from the string of equalities

$$\mathcal{A} = \lim_{n \to \infty} \mathcal{T}^n = \lim_{n \to \infty} \mathcal{T}^n \mathcal{T} = \mathcal{A}\mathcal{T} = \lim_{n \to \infty} \mathcal{T}\mathcal{T}^n = \mathcal{T}\mathcal{A}.$$

□

Thus, for a regular Markov chain, regardless of the initial vector $\mathbf{p}_0$, the long range—or *asymptotic*—probabilities of the states will be those specified by the vector $\alpha = (\alpha_1, \dots, \alpha_q)$. The vector α is often referred to as the *stationary distribution* of the Markov chain $(\mathbf{X}_n)$.

22.20. Example. The speed of the convergence $T^n \to \mathcal{A}$ in Theorems 22.17 and 22.19 is illustrated by the few powers of the matrix T displayed in Table 22.1. All the entries in T^3, T^{10}, and T^{40} are rounded to the third decimal.

Table 22.1: An exemplary transition matrix T and three of its powers T^n $(n = 3, 10, 40)$ in Example 22.20.

$$T = \begin{pmatrix} 0.6 & 0.3 & 0.1 & 0 & 0 \\ 0.8 & 0.1 & 0 & 0.1 & 0 \\ 0 & 0.1 & 0.7 & 0.2 & 0 \\ 0 & 0.1 & 0.1 & 0.6 & 0.2 \\ 0.1 & 0 & 0.1 & 0. & 0.8 \end{pmatrix} \qquad T^3 = \begin{pmatrix} 0.536 & 0.22 & 0.156 & 0.078 & 0.01 \\ 0.546 & 0.21 & 0.128 & 0.086 & 0.03 \\ 0.132 & 0.112 & 0.396 & 0.274 & 0.086 \\ 0.152 & 0.092 & 0.182 & 0.272 & 0.302 \\ 0.18 & 0.064 & 0.192 & 0.048 & 0.516 \end{pmatrix}$$

$$T^{10} = \begin{pmatrix} 0.384 & 0.170 & 0.203 & 0.141 & 0.102 \\ 0.384 & 0.169 & 0.203 & 0.140 & 0.104 \\ 0.305 & 0.141 & 0.222 & 0.158 & 0.174 \\ 0.307 & 0.139 & 0.217 & 0.144 & 0.193 \\ 0.308 & 0.138 & 0.218 & 0.138 & 0.198 \end{pmatrix}$$

$$T^{40} = \begin{pmatrix} 0.346 & 0.155 & 0.211 & 0.144 & 0.144 \\ 0.346 & 0.155 & 0.211 & 0.144 & 0.144 \\ 0.346 & 0.155 & 0.211 & 0.144 & 0.144 \\ 0.346 & 0.155 & 0.211 & 0.144 & 0.144 \\ 0.346 & 0.155 & 0.211 & 0.144 & 0.144 \end{pmatrix} \approx \mathcal{A}.$$

In the last section of this chapter, we apply Theorems 22.16, 22.17, and 22.19 to an example involving a stochastic model for the evolution of the preferences of individuals (as revealed, for instance, by opinion polls).

The Evolution of Preferences

22.21. Example. The states of the Markov chain track down the preferences of some individual regarding a set of alternatives—for instance, the candidates in a presidential election. Specifically, suppose that these preferences take the form of rankings of three candidates, without ties. Denoting the three candidates by the letters

b, g, and n, we get the six possible rankings

$$b \succ g \succ n, \quad b \succ n \succ g, \quad n \succ b \succ g,$$
$$n \succ g \succ b, \quad g \succ n \succ b, \quad g \succ b \succ n,$$

which form the six possible states of the chain. Thus, $b \succ g \succ n$ means: b is preferred to g, who is chosen over n. In the sequel, we often abbreviate $i \succ j \succ k$ into ijk, for i, j, and k in $\{b, g, n\}$. We assume that the current ranking of the individual may only change into another ranking that is 'adjacent' to it. For instance, bgn may only change into gbn or into bng, both of which can be obtained by transposing a single pair of bng. The directed graph of the chain is pictured in Fig. 22.2. Thus, state S can change into state T if there is an directed edge (or 'arc', see 23.11) linking T to S in the graph.

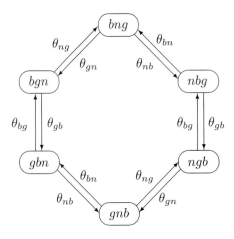

Figure 22.2: Transition diagram of the homogeneous Markov chain in the model of Example 22.21. As usual, the loops are omitted.

The θ_{ij}, with $i, j \in \{n, b, g\}$, flanking the edges of the graph denote the probabilities of the transitions between states. The transition matrix of the Markov chain is displayed in Table 22.2. The idea is that, under the influence of random, possibly unobservable events in the environment, the ranking of the three candidates by a potential voter may change by quantum jumps, each involving a transposition of a single pair of the preference order. In comprehensive presentations of this model given elsewhere (see for example Falmagne, 1997, Falmagne and Doignon, 1997, Regenwetter et al., 1999), the changes in the preferences are formalized by a stochastic process evolving in continuous time. Here, however, we only consider the Markov chain part of the model. We now turn to a formal discussion of this model.

Table 22.2: Transition matrix of the Markov chain in Example 22.21. As usual, all the rows in the matrix add up to 1. To save space in the display of the table, we omit the diagonal entries. For instance, the symbol ' $*$ ' in the first line should be replaced by ' $1 - \theta_{ng} - \theta_{gb}$.' A similar convention applies to the other lines.

	bgn	bng	nbg	ngb	gnb	gbn
bgn	$*$	θ_{ng}	0	0	0	θ_{gb}
bng	θ_{gn}	$*$	θ_{nb}	0	0	0
nbg	0	θ_{bn}	$*$	θ_{gb}	0	0
ngb	0	0	θ_{bg}	$*$	θ_{gn}	0
gnb	0	0	0	θ_{ng}	$*$	θ_{bn}
gbn	θ_{bg}	0	0	0	θ_{nb}	$*$

22.22. Definition. Let $(\mathbf{X}_n)$ be a sequence of jointly distributed random variables taking their values in the set

$$\mathcal{S} = \{bgn, bng, nbg, ngb, gnb, gbn\}$$

of all the rankings.[1] We assume that for any integer $n \geq 0$ and any sequence of states $S_0, \ldots S_n, S_{n+1}$, we have, with i, j, and k denoting distinct elements in $\{b, g, n\}$,

$$\mathbb{P}(\mathbf{X}_{n+1} = S_{n+1} \mid \mathbf{X}_n = S_n, \ldots, \mathbf{X}_0 = S_0)$$

$$= \begin{cases} \theta_{ij} & \text{if } S_n = jik \text{ and } S_{n+1} = ijk, \\ & \text{or } S_n = kji \text{ and } S_{n+1} = kij, \\ 1 - \theta_{ji} - \theta_{kj} & \text{if } S_n = S_{n+1} = ijk, \\ 0 & \text{in all other cases.} \end{cases} \quad (22.11)$$

Together with the initial distribution specifying the probabilities $\mathbb{P}(\mathbf{X}_0 = S)$ for any $S \in \mathcal{S}$, Eq. (22.11) specifies the sequence $(\mathbf{X}_n)$

---

[1]Random variables are real valued functions by definition. Technically, for the $\mathbf{X}_n$ to be random variables, we would have to change our notation of the elements of $\mathcal{S}$; for instance, we could number them 1, ... , 6. In this particular case, as the change would be trivial, we keep the more explicit notation of the rankings which makes the presentation easier to follow.

as a Markov chain with state space $\mathcal{S}$, whose transition diagram and transition matrix are given by Fig. 22.2 and Table 22.2, respectively.

22.23. Main Asymptotic Result. From the transition diagram, it is clear that the Markov chain is regular: Any state is accessible from any other state (cf. 22.4 and Theorem 22.16; in fact, this can be achieved in at most three steps). Applying Theorem 22.17, we know that the asymptotic probabilities of the states

$$\alpha(ijk) = \lim_{n\to\infty} \mathbb{P}(\mathbf{X}_n = ijk) \qquad (ijk \in \mathcal{S}) \qquad (22.12)$$

are defined. To determine those asymptotic probabilities, we proceed as follows. Using the abbreviations

$$p_n(ijk) = \mathbb{P}(\mathbf{X}_n = ijk) \qquad (n \geq 0,\ ijk \in \mathcal{S}),$$

we have by Theorem 9.8 (Total Probabilities) and Eq. (22.11), for any integer $n \geq 0$,

$$p_{n+1}(ijk) = \sum_{S \in \mathcal{S}} \mathbb{P}(\mathbf{X}_{n+1} = ijk \mid \mathbf{X}_n = S)\,\mathbb{P}(\mathbf{X}_n = S) \qquad (22.13)$$

$$= \theta_{ij}\, p_n(jik) + \theta_{jk}\, p_n(ikj) + (1 - \theta_{ji} - \theta_{kj})\, p_n(ijk),$$

because the other three terms in the right hand side of (22.13) vanish. Taking limits on both sides and using (22.12) yields

$$\alpha(ijk) = \theta_{ij}\alpha(jik) + \theta_{jk}\alpha(ijk) + (1 - \theta_{ji} - \theta_{kj})\alpha(ijk),$$

or, equivalently

$$\alpha(ijk)(\theta_{ji} + \theta_{kj}) = \theta_{ij}\alpha(jik) + \theta_{jk}\alpha(ijk). \qquad (22.14)$$

Equation (22.14) is the generic form for the six cases corresponding to the six ranking states in $\mathcal{S}$. Together with the equation

$$\sum_{S \in \mathcal{S}} \alpha(S) = 1, \qquad (22.15)$$

(22.14) defines a set of six independent equations in the unknowns $\alpha(ijk)$ and in the parameters θ_{ij}. The solution of this system, which is easily obtained, is

$$\alpha(ijk) = \frac{\theta_{ij}\theta_{ik}\theta_{jk}}{\Delta}. \qquad (22.16)$$

where

$$\Delta = \theta_{ij}\theta_{ik}\theta_{jk} + \theta_{ik}\theta_{ij}\theta_{kj} + \theta_{ji}\theta_{jk}\theta_{ik} + \theta_{jk}\theta_{ji}\theta_{ki} + \theta_{ki}\theta_{kj}\theta_{ij} + \theta_{kj}\theta_{ki}\theta_{ji}$$

is a normalizing factor ensuring that the $\alpha(ijk)$ values add up to 1. It is easily verified that the α values given by (22.16) satisfy (22.14).

The solution given by (22.16) can be rewritten in a particularly enlightening form. Let us write $\succ$ for any of the six rankings in $\mathcal{S}$, and define

$$\gamma(\succ) = \Pi_{i\succ j}\theta_{ij},$$

where Π denotes the usual product of real numbers. (Thus, in particular, $\gamma(bng) = \theta_{bn}\theta_{bg}\theta_{ng}$.) Then, Eq. (22.16) can be rewritten as

$$\alpha(\succ) = \frac{\gamma(\succ)}{\sum_{\succ' \in \mathcal{S}} \gamma(\succ')}. \tag{22.17}$$

In words: THE ASYMPTOTIC PROBABILITY OF STATE $\succ$ IS PRO-PORTIONAL TO THE PRODUCT OF ALL THE θ_{ij} SUCH THAT $i \succ j$. This result holds for any number of candidates that have to be ranked. In fact, it holds under essentially the same form for a vast class of models for the evolution of preferences (see Falmagne, 1997).

Problems

1. A set B of states is called *closed* if there is no state outside of B which is accessible from some state in B. Show that the intersection of closed sets of states is also closed. Accordingly, it makes sense to talk about the *smallest* (for the inclusion relation) closed set including a set B of states. This smallest closed set is the intersection of all the closed sets including B. The *closure* of a set of states is the smallest closed set including it. Since the set $\mathcal{S}$ of all states is closed by definition, the closure operation is always defined. Compute the closure of the sets of states $\{2,5\}$ and $\{4,5,6\}$ in Example 22.6.

2. (Continuation.) Prove that the closure of a set B of states is the set of all states accessible from some state in B.

3. Prove that if two states are mutually accessible, then they necessarily have the same period.

4. Compute explicitly the stationary distribution of a regular Markov chain with state space $\mathcal{S} = \{1,2,3\}$ and transition matrix $\mathcal{T} = (\tau(i,j))$. In the notation of this chapter, we have thus $t_{n,n+1}(i,j) = \tau(i,j)$.

5. Prove Eq. (22.17) in the general case of an arbitrary number of candidates to be ranked. (Thus $\alpha(\succ) = \lim_{n\to\infty} \mathbb{P}(\mathbf{X}_n = \succ)$, and $\succ$ denotes any of the $r!$ rankings of a set containing r candidates.)

6. Prove that (23.8) is indeed the solution of the system formed by the six equations defined by (22.14) and (22.15).

7. Suppose a homogeneous Markov chain with state space $\{1,\dots,6\}$ has transition matrix

$$\mathcal{T} = \begin{pmatrix} 0 & \frac{1}{3} & 0 & \frac{2}{3} & 0 & 0 \\ 0 & 0 & 1 & 0 & 0 & 0 \\ \frac{1}{2} & 0 & 0 & \frac{1}{2} & 0 & 0 \\ 0 & 0 & 0 & \frac{1}{2} & \frac{1}{4} & \frac{1}{4} \\ 0 & 0 & 0 & 0 & \frac{1}{3} & \frac{2}{3} \\ 0 & 0 & 0 & 0 & 1 & 0 \end{pmatrix}.$$

(a) Specify the set $\mathbb{S}$ of classes of states. (b) Define the partial order $\precsim$ on $\mathbb{S}$ corresponding to $\rightarrowtail$, and identify each class as transient or ergodic. Are there any absorbing states? (c) For each state j, give the probability $f(j)$ of ever returning to that state.

8. Prove that if two states are mutually accessible and one is persistent, then the other also is persistent.

9. What can be said about the closure of an ergodic class? Of a transient class?

Chapter 23

Random Walks

The Markovian model for the evolution of preferences discussed at the end of Chapter 22 had the remarkable property that transitions could only happen between 'neighbor' states.[1] This type of transitions arises in an important class of Markov chains called 'random walks.'[2] A key difference between the model of Example 22.21 and those discussed here is that the models of this chapter have their state space countable rather than finite.

The coin tossing game encountered in 21.4, which involves the two players Peter and Paul, was suggestive of such random walks. In this game, Peter gives \$1 to Paul in case of a head and receives \$1 in the other case. We assume that the probability of a head is equal to γ, constant over trials. We keep track of the gain of Peter, which fluctuates—until the ruin of one of the two players—between -\$100 and \$100. Suppose that we measure the gain of Peter on trial n by a random variable $\mathbf{X}_n$, $n = 0, 1, \dots$. Then, the sequence $(\mathbf{X}_n)$ is a Markov chain with state space $\{-100, -99, \dots, 0, \dots, 100\}$ and initial probability distribution defined by $\mathbb{P}(\mathbf{X}_0 = 0) = 1$.

In the next section, we generalize this situation in two ways: (1) We assume that the coin could fall on its edge so that the gain of Peter may remain constant on some trials; and (2) we suppose that the fortune of the two players is infinite; the state space is thus countable, consisting of the set of all integers.

---

[1] In this particular case, every state has exactly two such neighbors. For instance, the only transitions from state *bgn* are to states *gbn* and *bng* in Example 22.21.

[2] Some authors would refer to the Markov chain of Example 22.21 and Def. 22.22 as a 'circular random walk' (for example, Feller, 1957).

A Random Walk on the Integers

23.1. Definition. Let $(\xi_n)_{n\in\mathbb{N}_0}$ be a sequence of functions, with $\mathbb{N}_0 = \{0, 1, \dots\}$ and $\xi_n : \mathbb{Z} \times \mathbb{Z} \to [0, \infty[$ for $n \in \mathbb{N}_0$. Let $(\mathbf{X}_n)_{n\in\mathbb{N}_0}$ be a sequence of random variables satisfying the two following conditions:

[W1] $\mathbb{P}(\mathbf{X}_0 = 0) = 1$;

[W2] For any integer $n \geq 0$,

$$\mathbb{P}(\mathbf{X}_{n+1} = j \mid \mathbf{X}_n = i, \dots, \mathbf{X}_1, \mathbf{X}_0) = \mathbb{P}(\mathbf{X}_{n+1} = j \mid \mathbf{X}_n = i) \quad (23.1)$$
$$= \xi_n(i, j).$$

Then the sequence $(\mathbf{X}_n)$ is called a *Markov chain* on $\mathbb{Z}$ with *transition functions* ξ_n. As in the finite case, the set $\mathbb{Z}$ is referred to as the *state space*. If the functions ξ_n do not depend on the index n, the Markov chain $(\mathbf{X}_n)$ is said to be *homogeneous*[3] with *transition function* $\xi = \xi_n$ (for all $n \in \mathbb{N}_0$). Consider the additional axiom:

[W3] The Markov chain $(\mathbf{X}_n)$ is homogeneous with transition function

$$\xi(i, j) = \begin{cases} \theta & \text{if } j = i + 1, \\ \gamma & \text{if } j = i - 1, \\ 1 - \theta - \gamma & \text{if } j = i, \\ 0 & \text{in all other cases.} \end{cases} \quad (23.2)$$

Then, the sequence of random variables $(\mathbf{X}_n)$ is called a *(homogeneous) random walk* on $\mathbb{Z}$. Note that the sequence $(\mathbf{X}_n)$ is still referred to as a Markov chain (without the qualifier 'finite').

23.2. Remarks. (a) Equation (23.1) is exactly that used to define Markov chains in the finite case (see Eqs. (21.4), (21.5) and Problem 5 in Chapter 21). Thus, random walks on the integers generalize a class of finite Markov chains in which every state in a countable state space has exactly two neighbors and the only possible transitions occur between neighbors.

(b) Notice that, by contrast with the coin tossing game between Peter and Paul, [W3] has no stopping rule. In Problem 1, we ask the reader to rewrite the right-hand side of (23.2) so as to match the situation of Example 21.4.

---

[3]Note that this terminology is consistent with Def. 22.1.

We now give another empirical application of the concept of a random walk which is rather different from that of the coin tossing game and is suggestive of generalizations.

23.3. Example. Axioms [W1], [W2], and [W3] formalize a possible mechanism governing the movements of a particle on the real line. In this framework, the particle can only occupy integral positions on that line. Suppose that we orient the line vertically, with the time running horizontally. We assume that the particle can only move one quantum step at the time, either up or down, or stay in place. These movements occur only at discrete instants in time. The probabilities of the up and down movements are θ and γ, respectively; with probability $1 - \theta - \gamma$ that no move occurs (cf. Axiom [W3]). An exemplary realization[4] of the process is displayed in Fig. 23.1.

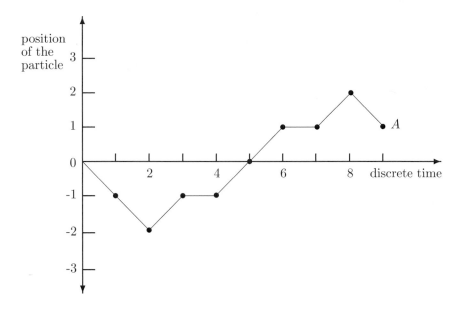

Figure 23.1: Exemplary realization of a random walk on the set of integers (see Definition 23.1). With θ and ξ as in Axiom [W3], the probability of the particular path displayed, from the origin to point A, is $\gamma^3(1 - \gamma - \theta)^2\theta^4$.

---

[4]In the theory of Markov chains, and more generally, of stochastic processes evolving in discrete time, a feasible sequence of values of the random variables may be called a *realization* of the process.

23.4. Remarks. Numerous variants of Axioms [W1], [W2], and [W3] are available which may also define random walks. We only describe some of them informally here.

(a) Suppose that we modify Axiom [W3] so that the transition probabilities may depend on the trial number n, that is, we replace θ and γ in the right hand side of (23.2) by θ_n and γ_n. We still have a random walk, but it is no longer homogeneous.

(b) By extension, the coin tossing game of Example 21.4 between Peter and Paul, in which a sequence of random variables keeps track of Peter's gain, is also regarded as an homogeneous random walk with two 'absorbing barriers' 100 and -100 corresponding to the ruin of each of the two players. The distinction is academic because, in the framework of a homogeneous random walk in the sense of Def. 23.1, we can compute the probability of such events as 'the random walk hits 100 before hitting -100.'

(c) We may take $\mathbb{N}_0$ as the state space, rather than $\mathbb{Z}$. Axiom [W1] remains, and so does [W2], except that the functions ξ_n of [W2] are now mapping $\mathbb{N}_0 \times \mathbb{N}_0$ into $\mathbb{N}_0$. State $0 \in \mathbb{N}_0$ has a single neighbor. All the other states $i \in \mathbb{N}_0 \setminus \{0\}$ have two neighbors, $i - 1$ and $i + 1$. Suppose that the random walk is homogeneous, with a transition matrix as displayed in Table 23.1, infinite in both directions.

Table 23.1: Transition matrix of the random walk on $\mathbb{N}_0$ in 23.4(b), infinite in both directions. We suppose that $0 < \gamma < 1 - \theta$. If the random walk hits 0 at any time, it bounces back on the next trial with probability θ.

	0	1	2	3	4	...
0	$1-\theta$	θ	0	0	0	...
1	γ	$1-\theta-\gamma$	θ	0	0	...
2	0	γ	$1-\theta-\gamma$	θ	0	...
3	0	0	γ	$1-\theta-\gamma$	θ	...
4	0	0	0	γ	$1-\theta-\gamma$	...
⋮	⋮	⋮	⋮	⋮	⋮	...

The reader should compare this transition matrix with that of Table 22.2, and also with Eq. (23.2). Note that if the random walk hits 0 at any time, it may bounce back up on the next trial with probability θ. State 0 is called a *reflecting barrier* of the random walk. We could modify this Markov chain so that, if it hits 0 at any time after the first trial, then it would forever remain at 0. In this modified situation, state 0 is an absorbing state (cf. 22.7) which is referred to as an *absorbing barrier* in the theory of random walks.

23.5. Convention. In the next section, we focus on the random walk defined by Axioms [W1], [W2] and [W3] of Def. 23.1, but with $\theta + \gamma = 1$. Occasionally, it will be convenient and intuitively helpful to adopt the language of Example 23.3 and to say, for example, for some integers $r, q > 0$,

'THE PARTICLE WILL HIT r BEFORE HITTING $-q$' (23.3)

to mean the event

$$\cup_{n=r}^{\infty} \left(\{\omega \in \Omega \mid \mathbf{X}_n(\omega) = r\} \cap_{j=0}^{n-1} \{\omega \in \Omega \mid -q < \mathbf{X}_j(\omega) < r\} \right) \quad (23.4)$$

where Ω denotes the sample space, that is, in this case, the set of all sequences in $\mathbb{Z}$. To understand that (23.4) is a valid translation of (23.3), one must see: First, that the particle may hit r on any trial $n \geq r$, hence the union from $n = r$ to ∞ in (23.4); and second, that among all those cases of such a hit for a particular value of $n \geq r$, we must retain only those subcases where there were no earlier hits of either $-q$ or r, hence the intersection from $j = 0$ to $n - 1$ in (23.4), which effectively removes all the subcases of earlier hits.

Needless to say, the reader must be able to go back and forth from the 'particle' language to the formal expressions, which may be useful for the computation of some probabilities (see Problem 3 in this connection).

The Probability of Hitting One of Two Bounds

23.6. Problem. In the case $\theta + \gamma = 1$, find the probability that the particle, starting at $k > 0$, hits 0 before hitting r, where r is any positive integer such that $r > k$.

The rest of this chapter is devoted to solving this problem and related ones. (We follow Feller, 1957, who gives an elementary discussion. For more details, see also Parzen, 1994b, or Barucha-Reid, 1974.)

23.7. Remarks. (a) On the face of it, Problem 23.6 is a special case of the following one: 'Let $q < j < t$ be any integers. In the case $\theta + \gamma = 1$, find the probability that the particle, starting at j, hits q before hitting t.' Actually, the extra generality of this second problem is superficial. In both problems, we are dealing with the movement of a particle along the integer line, with two probabilities θ and γ of moving up and down, respectively. The second version can be recast as the first one by 'sliding' the interval $[q, t]$ appropriately and setting $k = j - q > 0$ and $r = t - q > k$.

(b) In the same vein, note also the symmetry between Problem 23.6 and the problem of computing the probability that the particle hits the upper bound r before the lower bound 0. When a formula F solving Problem 23.6 has been found, the probability that the particle hits r before 0 can be computed simply by interchanging, in F, γ and θ on the one hand, and k and $r - k$ on the other hand. This will generate a formula F^* that will satisfy $F + F^* = 1$, establishing the fact that the event 'the particle never reaches either 0 or r' has measure 0 (cf. Eq. (23.17)).

23.8. Solution of Problem 23.6. It is expedient to interpret the statement 'starting at $k > 0$' in Problem 23.6 by recasting Axiom [W1] as

[W1'] $\mathbb{P}(\mathbf{X}_0 = k) = 1.$

The Markovian Axiom [W2] of Def. 23.1 is unchanged, and [W3] is specialized by assuming that $\theta + \gamma = 1$. In other words, with probability 1, the particle moves up or down on every trial. The problem asks for the probability of the event

$$E_k = \cup_{n=k}^{\infty} \left(\{\omega \in \Omega \mid \mathbf{X}_n(\omega) = 0\} \cap_{j=1}^{n-1} \{\omega \in \Omega \mid 0 < \mathbf{X}_j(\omega) < r\} \right).$$

We adopt the abbreviation

$$\nu_k = \mathbb{P}(E_k).$$

On the first trial, the particle hits either $k+1$ or $k-1$, with respective probabilities θ and γ. Using the Theorem of Total Probabilities (cf. 9.8), we get for $0 < k < r$

$$\begin{aligned} \nu_k &= \mathbb{P}(\mathbf{X}_1 = k + 1)\mathbb{P}(E_k \mid \mathbf{X}_1 = k + 1) \\ &\quad + \mathbb{P}(\mathbf{X}_1 = k - 1)\mathbb{P}(E_k \mid \mathbf{X}_1 = k - 1) \\ &= \theta\,\mathbb{P}(E_k \mid \mathbf{X}_1 = k + 1) + \gamma\,\mathbb{P}(E_k \mid \mathbf{X}_1 = k - 1). \end{aligned} \qquad (23.5)$$

Suppose that $1 < k < r - 1$. In this case, the conditional probability $\mathbb{P}(E_k \,|\, \mathbf{X}_1 = k + 1)$ is essentially the probability that the particle returns to 0 before hitting r if it starts at $k+1$, so that we can write

$$v_{k+1} = \mathbb{P}(E_k \,|\, \mathbf{X}_1 = k + 1).$$

For a similar reason, we can write $v_{k-1} = \mathbb{P}(E_k \,|\, \mathbf{X}_1 = k - 1)$, so that, for $1 < k < r - 1$, (23.5) becomes

$$v_k = \theta \, v_{k+1} + \gamma \, v_{k-1}. \tag{23.6}$$

If $k = 1$ or $k = r - 1$, the particle may hit 0 or r, respectively, on the first trial. In those cases, (23.6) must be replaced by

$$v_1 = \theta v_2 + \gamma \tag{23.7}$$

$$v_{r-1} = \gamma v_{r-2}. \tag{23.8}$$

To gather (23.7) and (23.8) in the framework of (23.6), we adopt the conventions

$$v_0 = 1 \quad \text{and} \quad v_r = 0, \tag{23.9}$$

which are sensible 'boundary conditions' in the context of our problem. With these conventions, Eq. (23.6) holds for $1 \le k \le r - 1$.

CASE 1. We begin by assuming that $\theta \ne \gamma$. Notice that, for any constants c and d, the form

$$v_k = c + d \, (\gamma/\theta)^k \tag{23.10}$$

satisfies (23.6) (taken by itself, that is, without requiring the boundary conditions (23.9)). Indeed, replacing v_{k+1} and v_{k-1} in (23.6) by their expressions according to (23.10) and rearranging yields

$$
\begin{aligned}
v_k &= \theta v_{k+1} + \gamma v_{k-1} \\
&= \theta \left(c + d \, (\gamma/\theta)^{k+1} \right) + \gamma \left(c + d \, (\gamma/\theta)^{k-1} \right) \\
&= (\theta + \gamma) c + d \left(\theta \, (\gamma/\theta)^{k+1} + \gamma \, (\gamma/\theta)^{k-1} \right) \\
&= (\theta + \gamma) c + (\gamma + \theta) d \, (\gamma/\theta)^k \\
&= c + d \, (\gamma/\theta)^k .
\end{aligned}
$$

Our boundary conditions (23.9), however, fix the values of the constants c and d, which must satisfy the two equations

$$1 = v_0 = c + d(\gamma/\theta)^0, \qquad 0 = v_r = c + d(\gamma/\theta)^r.$$

This yields

$$c = \frac{(\gamma/\theta)^r}{(\gamma/\theta)^r - 1},$$ (23.11)

$$d = \frac{-1}{(\gamma/\theta)^r - 1}.$$ (23.12)

Replacing c and d in (23.10) by their expressions in (23.11) and (23.12) gives the following solution to Problem 23.6:

$$\mathbb{P}(E_k) = v_k = \frac{(\gamma/\theta)^r - (\gamma/\theta)^k}{(\gamma/\theta)^r - 1}.$$ (23.13)

This solution is unique. To prove this, notice that Eq. (23.6) specifies v_{k+1} in terms of v_k and v_{k-1}. The values of v_1 and v_2 thus suffice to determine v_k for any $3 \leq k \leq r - 1$, and so two solutions of Problem 23.6 giving the same values for v_1 and v_2 must be identical. Now suppose that $f(k) = v_k$ is any solution of (23.6) satisfying the boundary conditions (23.9). By choosing the values of c and d appropriately, we can always ensure that $f(1) = c + d(\gamma/\theta)$ and $f(2) = c + d(\gamma/\theta)^2$, which implies, by the above argument and a simple induction, that

$$f(k) = c + d(\gamma/\theta)^k \qquad (1 \leq k \leq r - 1).$$ (23.14)

However, the function f in (23.14) must verify (23.7) and (23.8) with $f(k) = v_k$. As is easily shown, this forces the chosen values of c and d to satisfy (23.11) and (23.12), and so (23.13) also holds in this case with $f(k) = v_k$. We conclude that (23.13) is the unique solution of Problem 23.6 in Case 1.

CASE 2. If $\theta = \gamma = \frac{1}{2}$, Equation (23.13) becomes meaningless (however, see Problem 5). We can easily verify that, in this case, the form $v_z = c + dz$ verifies (23.6). As we still have the boundary conditions (23.9), we get $v_0 = c = 1$ and $v_r = c + dr = 1 + dr = 0$. Thus $d = -\frac{1}{r}$. This yields

$$v_k = 1 - \frac{k}{r}$$ (23.15)

as the solution in Case 2. We leave to the student the verification that this second solution is also unique (Problem 6).

In connection with Remark 23.7(b), we consider now the problem of determining the probability w_k that the particle, starting at k, hits r before 0. This problem and Problem 23.6 are clearly symmetric to each other. In Case 1, (that is, $\theta \neq \gamma$), we can compute w_k by a simple manipulation of Eq. (23.13) interchanging the roles of the quantities k and $r - k$ on the one hand, and θ and γ on the other hand. This gives for the probability that the particle hits r before k the expression

$$w_k = \frac{(\theta/\gamma)^r - (\theta/\gamma)^{r-k}}{(\theta/\gamma)^r - 1} = \frac{(\gamma/\theta)^k - 1}{(\gamma/\theta)^r - 1}. \qquad (23.16)$$

Comparing (23.16) and (23.13), we see that

$$v_k + w_k = 1, \qquad (23.17)$$

which means that, in Case 1, the particle will hit one of the two bounds with probability 1. The same result is obtained in Case 2.

We summarize our solution of Problem 23.6 as follows.

A PARTICLE STARTING IN POSITION $k > 0$ HAS A PROBABILITY v_k OF HITTING 0 BEFORE r SPECIFIED BY THE EQUATIONS:

$$v_k = \begin{cases} \frac{(\gamma/\theta)^r - (\gamma/\theta)^k}{(\gamma/\theta)^r - 1} & \text{if } \theta \neq \gamma, \\ 1 - \frac{k}{r} & \text{if } \theta = \gamma = \frac{1}{2}. \end{cases} \qquad (23.18)$$

THE PROBABILITY THAT THE PARTICLE NEVER HITS ONE OF THE TWO BOUNDS 0 AND r IS EQUAL TO ZERO.

It is also of interest to rephrase these conclusions in the language of the game situation described in 21.4 (see Problem 7).

Expected Time for Hitting a Bound

23.9. Problem. In the situation of Problem 23.6, suppose that the particle starts at k. What is the expected time τ_k for the particle to hit either of the two bounds 0 and r, with $0 < k < r$? (We have thus $\theta + \gamma = 1$.)

We are lacking some of the equipment required to deal with this problem in its full generality. To simplify matters, we shall assume without proof that the expectation τ_k is finite. (This is true.)

23.10. Solution of Problem 23.9. Thus, τ_k stands for the conditional expectation of the 'time for the particle to hits either one of the two bounds 0 and r' given the event 'the particle starts at k.' As in 23.6, the sequence $(\mathbf{X}_n)_{n \in \mathbb{N}_0}$ measures the position of the particle on the successive trials, and we also replace Axiom [W1] by [W1'], which states that $\mathbb{P}(\mathbf{X}_0 = k) = 1$. Denoting by $\mathbf{T}$ the random variable 'time to hit one of the two bounds' we get $\tau_k = E(\mathbf{T})$. On the first trial, either the particle moves toward 0 or toward r by one step. From the properties of conditional expectation, we obtain

$$\tau_k = \theta E(\mathbf{T} \mid \mathbf{X}_1 = k + 1) + \gamma E(\mathbf{T} \mid \mathbf{X}_1 = k - 1). \qquad (23.19)$$

Now, $E(\mathbf{T} \mid \mathbf{X}_1 = k + 1)$ is the expected time, counted from trial 0, for the particle to reach a bound, given that on trial 1, the particle is at $k + 1$. This means that we can write

$$E(\mathbf{T} \mid \mathbf{X}_1 = k + 1) = \tau_{k+1} + 1,$$

and for symmetrical reasons

$$E(\mathbf{T} \mid \mathbf{X}_1 = k - 1) = \tau_{k-1} + 1,$$

and so (23.19) gives

$$\tau_k = \theta \tau_{k+1} + \gamma \tau_{k-1} + 1 \qquad (0 < k < r) \qquad (23.20)$$

which resembles (23.6) but is nonhomogenous. In the spirit of the last section, we adopt the natural boundary conditions

$$\tau_0 = \tau_r = 0. \qquad (23.21)$$

As in the solution of Problem 23.6, we have two cases.

CASE 1. Assume that $\theta \neq \gamma$. It can be checked that

$$\tau_k = \frac{k}{\gamma - \theta}. \qquad (23.22)$$

is a solution of (23.20). Suppose now that Eq. (23.20) has two distinct solutions $\zeta_{1,k}$ and $\zeta_{2,k}$ and define

$$\delta_k = \zeta_{1,k} - \zeta_{2,k}. \qquad (23.23)$$

Thus, each of $\zeta_{1,k}$ and $\zeta_{2,k}$ satisfies (23.20) and we obtain

$$\delta_k = \theta \delta_{k+1} + \gamma \delta_{k-1}. \qquad (23.24)$$

This equation is of the form of (23.6), and we know from our discussion of Case 1 in 23.8 that, if $\theta \neq \gamma$, all the solutions of (23.24) are of the form

$$\delta_k = c + d \left(\gamma/\theta\right)^k \tag{23.25}$$

for some constants c and d. Taking $\zeta_{2,k} = k/(\gamma - \theta)$ in (23.23), we obtain from (23.25)

$$c + d \left(\gamma/\theta\right)^k = \zeta_{1,k} - \frac{k}{\gamma - \theta},$$

for any solution $\zeta_{1,k}$ of (23.20). This implies that the general form of (23.20) must be

$$\tau_k = \frac{k}{\gamma - \theta} + c + d \left(\gamma/\theta\right)^k. \tag{23.26}$$

Because this equation must satisfy the boundary conditions (23.21), we get the two equations $0 = c + d$ and $0 = r/(\gamma - \theta) + c + d \left(\gamma/\theta\right)^r$. Replacing c and d in (23.26) by the solutions of this system gives finally

$$\tau_k = \frac{k}{\gamma - \theta} - \frac{r}{\gamma - \theta} \cdot \frac{1 - (\gamma/\theta)^k}{1 - (\gamma/\theta)^r} \tag{23.27}$$

as the general solution in Case 1 of Problem 23.9.

CASE 2. Suppose that $\theta = \gamma = \frac{1}{2}$. Notice that

$$\tau_k = -(k^2) \tag{23.28}$$

is then a solution of (23.20). We use an argument similar to that of Case 1, but with (23.28) playing here the role of $\tau_k = k/(\gamma - \theta)$ there. All the solutions of (23.20) must be of the form

$$\tau_k = -(k^2) + c + dk. \tag{23.29}$$

Plugging the boundary conditions (23.21) into (23.29) fixes the values of c and d. This yields

$$\tau_k = k(r - k). \tag{23.30}$$

Summing up the results for Problem 23.9, we get:

A PARTICLE STARTING IN POSITION $k > 0$ HAS AN EXPECTED TIME FOR REACHING EITHER ONE OF THE TWO BOUNDS 0 AND r SPECIFIED BY THE EQUATIONS:

$$v_k = \begin{cases} \frac{k}{\gamma-\theta} - \frac{r}{\gamma-\theta} \cdot \frac{1-(\gamma/\theta)^k}{1-(\gamma/\theta)^r} & \text{if } \theta \neq \gamma, \\ k(r-k) & \text{if } \theta = \gamma = \frac{1}{2}. \end{cases} \tag{23.31}$$

Some implications of these results are considered in Problem 8.

Generalization

The concept of random walk discussed in the previous section has a natural generalization in two or more dimensions. Figure 23.2 illustrates this idea by displaying a realization of a random walk on a two-dimensional grid. Every point of the grid represents a state of the random walk and a possible position of the particle. On each trial, the particle can either stay in place or move up, down, left or right. This situation can be axiomatized along lines similar to those of the random walk on the integers discussed earlier in this chapter (see Problem 9).

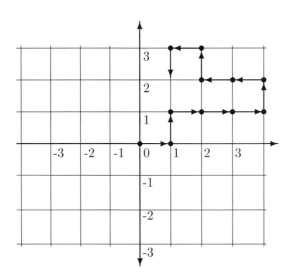

Figure 23.2: Exemplary realization of a random walk on a planar grid.

A Random Walk on a Connected Graph*

More general types of random walks can be conceived. As illustrated by our examples, a random walk is a Markov chain in a case where a nontrivial neighborhood relation is defined on the set of states, and the only possible transitions are those occuring between neighbor states. This suggests that, in principle, random walks can be constructed on the set of 'vertices' of a 'connected graph.' The analysis of such general random walks is difficult, however. We examine briefly here a case in which the machinery introduced in this chapter is nevertheless applicable for some purposes. We have in mind a situation in which the movement of a particle begins at a certain point of the connected graph, called the 'center,' and we are only interested in the expected time required for the particle to reach any point at a certain distance from the center. The result is easy to obtain if the connected graph satisfies certain conditions.

We begin with some relevant definitions.

23.11. Definition. In graph theory, a binary relation D on a set $\mathcal{S}$ is called a *directed graph* or *digraph* on $\mathcal{S}$. The points of $\mathcal{S}$ are referred to as the *vertices* of the digraph D and each pair $xy \in D$ is called a *directed edge* or *arc*. This language is consistent with a geometrical representation in which the vertices of the digraph are points and the arcs are pictured as arrows linking the points. An example is displayed in Figure 23.3 (a), with

$$\mathcal{S} = \{a, b, c, d\} \quad \text{and} \quad D = \{ab, ba, bc, cd\}.$$

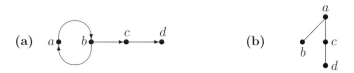

Figure 23.3: (a) A digraph on the set $\{a, b, c, d\}$; the corresponding binary relation is $\{ab, ba, bc, cd\}$. Thus, each of ab, ba, bc and cd is an arc of the graph.
(b) A graph on the same set $\{a, b, c, d\}$. The set of edges is the family of subsets $\{\{a, b\}, \{a, c\}, \{a, d\}\}$.

The adjective 'directed' is dropped when the relation D is symmetric, that is, when $xDy \Leftrightarrow yDx$, for all $x, y \in \mathcal{S}$. In this case, we simplify the writing and recode D as the family $G = \{\{x, y\} \mid xDy\}$.

Thus, G is a *graph* on a set S if G is a family of two-element subsets of S. The points of S are still called *vertices*. Each of these doubleton subsets is referred to as an *edge* of G. We also define, for any vertex x, the set $G_x = \{y \in S \mid \{x, y\} \in G\}$ of its *neighbors*. It is clear that $y \in G_x \Leftrightarrow x \in G_y$. An example is given in Figure 23.3 (b) with

$$S = \{a, b, c, d\}, \qquad G = \{\{a, b\}, \{a, c\}, \{c, d\}\}$$
$$G_a = \{b, c\}, \quad G_b = \{a\}, \quad G_c = \{a, d\}, \quad G_d = \{c\}.$$

23.12. Definition. A *path* connecting any two vertices $x, y \in S$ of a graph is a sequence of edges

$$\{x_0, x_1\}, \{x_1, x_2\}, \dots, \{x_{n-1}, x_n\} \tag{23.32}$$

linking $x = x_0$ to $y = x_n$, such that all the vertices x_i, $0 \le i \le n$ are distinct, except possibly x_0 and x_n. A graph G is called *connected* when, for any two distinct vertices x and y, there is a path connecting them. The graph pictured in Figure 23.3 (b) is connected.

In the rest of this section, we consider a connected graph G on a countable set set S. In addition, we suppose that G is *discrete*, that is, G satisfies the two following conditions:

1. any vertex $x \in S$ has a finite set of neighbors;

2. for any two vertices, there is a finite path connecting them.

In this case, we can define the *distance* $d(x, y)$ between two vertices x and y as the length of the shortest path between them; thus, $d(x, y) = d(y, x)$ for all vertices x and y. If the path given in (23.32) is the shortest one between x and y, we have $d(x, y) = n$.

A connected, discrete graph G satisfies the *even-triangle* condition if for any three distinct vertices x, y, and z, we have

$$d(x, y) + d(y, z) + d(z, x) = 2k$$

for some positive integer k.

23.13. Example. The planar grid of Figure 23.2 can be cast as a connected, discrete graph satisfying the even-triangle property. Each point of the grid is a vertex, and the edges have the two forms

$$\{(i, j), (i + 1, j)\} \quad \text{or} \quad \{(i, j), (i, j + 1)\}.$$

All the properties are easily verified (see Problem 11).

23.14. Theorem. *Suppose that x and y are neighbor vertices in a connected, discrete graph satisfying the even-triangle property. Then, for any vertex z, we have*

$$|d(z, x) - d(z, y)| = 1. \tag{23.33}$$

PROOF. Since x and y are neighbors, we have $d(x, y) = 1$. If z coincides with either x or y, the result follows trivially. So, suppose that $z \neq x$ and $z \neq y$. If $d(z, x) = d(z, y) = k$ for some $k \in \mathbb{N}$, we get $d(x, z) + d(z, y) + d(y, x) = 2k + 1$, contradicting the even-triangle property. Thus, we have necessarily $d(z, x) \neq d(z, y)$. We cannot have $|d(z, x) - d(z, y)| \geq 2$ because there would exist either a path of length $d(z, x) + d(x, y) < d(z, y)$ connecting z to y, or a path of length $d(z, y) + d(x, y) < d(z, x)$ connecting z to x, contradicting the minimality of $d(z, x)$ and $d(z, y)$. Equation (23.33) is the only remaining possibility. ☐

23.15. Remark. In the situation of Theorem 23.14, suppose that we fix some particular vertex C. Then, for any vertex x and any neighbor y of x, we have

$$\text{either} \qquad d(C, y) = d(C, x) + 1 \tag{23.34}$$
$$\text{or} \qquad d(C, y) = d(C, x) - 1. \tag{23.35}$$

23.16. Definition. Let $\mathcal{S}$ be the countable set of vertices of a discrete, connected graph G satisfying the even-triangle property. Fix a particular vertex C called the *center* of the graph. For every vertex $x \in \mathcal{S}$, define

$$G_x^+ = \{y \in G_x \mid d(C, y) = d(C, x) + 1\}$$
$$G_x^- = \{y \in G_x \mid d(C, y) = d(C, x) - 1\}.$$

By Remark 23.16, we have $G_x = G_x^+ + G_x^-$.

We now define a random walk on $\mathcal{S}$. Let θ and γ be two parameters, with $0 < \theta, \gamma$ and $\theta + \gamma \leq 1$, and let $\xi : \mathcal{S} \times \mathcal{S} \to [0, \infty[$ be a function defined, for $x \neq C$, by the equation

$$\xi(x, y) = \begin{cases} \dfrac{\theta}{|G_x^+|} & \text{if } y \in G_x^+, \\[2mm] \dfrac{\gamma}{|G_x^-|} & \text{if } y \in G_x^-, \\[2mm] 1 - \left(\dfrac{\theta}{|G_x^+|} + \dfrac{\gamma}{|G_x^-|} \right) & \text{if } x = y, \\[2mm] 0 & \text{in all other cases}. \end{cases} \tag{23.36}$$

For $x = C$, we define

$$\xi(C, y) = \begin{cases} \frac{\theta}{|G_C^+|} & \text{if } y \in G_C^+, \\ 1 - \frac{\theta}{|G_x^+|} & \text{if } y = C, \\ 0 & \text{in all other cases}. \end{cases} \qquad (23.37)$$

Let $(\mathbf{X}_n)_{n \in \mathbb{N}_0}$ be a sequence of random variable[5] taking their values on the set $\mathcal{S}$ of vertices of G. We suppose that the sequence $(\mathbf{X}_n)$ satisfies the two conditions:

[G1] $\mathbb{P}(\mathbf{X}_0 = C) = 1$;

[G2] For any integer $n \geq 0$ and any vertices x, y,

$$\mathbb{P}(\mathbf{X}_{n+1} = y \,|\, \mathbf{X}_n = x, \mathbf{X}_{n-1}, \dots, \mathbf{X}_1, \mathbf{X}_0) = \mathbb{P}(\mathbf{X}_{n+1} = y \,|\, \mathbf{X}_n = x)$$
$$= \xi(x, y),$$

with ξ defined by (23.36) and (23.36).

For the remainder of this section, an examination of the example of Figure 23.4 will be useful. The figure depicts part of a connected, discrete graph satisfying the even-triangle property and centered at the vertex C. All the vertices at distance ≤ 5 are displayed. In our discussion, we refer to this example as [EG].

Axiom [G1] ensures that the random walk begins at the center C. Because $G_C^- = \emptyset$, we have by [G2]

$$\mathbb{P}(\mathbf{X}_2 \in G_C^+ \,|\, \mathbf{X}_1 = C) = \mathbb{P}(\mathbf{X}_2 \in G_C^+) = \theta,$$
$$\mathbb{P}(\mathbf{X}_2 = C \,|\, \mathbf{X}_1 = C) = \mathbb{P}(\mathbf{X}_2 = C) = 1 - \theta.$$

Thus, with probability θ, the particle moves to a vertex at distance 1 from C on the first step. In [EG], the particle moves to one of $a_1, \dots, a_5$ with probability θ and remains at C with probability $1 - \theta$. Axiom [G2] implies that, on each step after that, the particle moves outward with probability θ, inward with probability γ and remains in place with probability $1 - \theta - \gamma$. In [EG], a particle located on vertex a_{10} on any trial n moves outward on trial $n + 1$ to one of a_{16}, a_{17} or a_{11}, with probability $\theta/3$ in each case. The particle moves inward to a_5 with probability γ, and remains on vertex a_{10}.

---

[5]As earlier in similar cases, we call the $\mathbf{X}_n$ random variables even though they are not technically so. (These functions are not real valued.) The distinction is immaterial here because the set $\mathcal{S}$ is countable.

Thus, if one is only interested in keeping track of the outward or inward moves of the particle, the analysis can be considerably simplified by projecting the random walk $(\mathbf{X}_n)$ on the nonnegative integers, which results in a random walk on that set, a case studied earlier in this chapter (see for example 23.6, 23.7 and 23.8). For example, we can easily solve the following problems.

23.17. Problems. Let $\mathcal{B}_r$ be the set of all the vertices at distance r from the center. Assume that $\theta + \gamma = 1$ (thus, with probability 1, the particle moves either outward or inward on any trial). Suppose that the particle is at distance $k < r$ on some trial.

(a) What is the probability that the particle returns to C before hitting $\mathcal{B}_r$?

(b) What is the expected time required for a hit of either $\mathcal{B}_r$ or C?

(c) What is the expected time required for a first hit of the set $\mathcal{B}_r$, given that it hits $\mathcal{B}_r$ before hitting C?

We leave the solution to the student (see Problem 14). The projection of the random walk $(\mathbf{X}_n)_{n \in \mathbb{N}_0}$ on the nonnegative integer is achieved via the next definition.

23.18. Definition. Let $(\mathbf{Y}_n)_{n \in \mathbb{N}_0}$ be a sequence of random variables defined from the random variables $\mathbf{X}_n$ by the equation

$$\mathbf{Y}_n = d(C, \mathbf{X}_n), \qquad (23.38)$$

where d is the distance function defined in 23.12.

23.19. Theorem. *The sequence of random variables* $(\mathbf{Y}_n)_{n \in \mathbb{N}_0}$ *is a random walk on* $\mathbb{N}_0$, *satisfying the two conditions*

(i) $\mathbb{P}(\mathbf{Y}_0 = 0) = 1$;

(ii) *For any integer* $n \geq 0$,

$$\mathbb{P}(\mathbf{Y}_{n+1} = j \mid \mathbf{Y}_n = i, \dots, \mathbf{Y}_1, \mathbf{Y}_0) = \mathbb{P}(\mathbf{Y}_{n+1} = j \mid \mathbf{Y}_n = i)$$
$$= \begin{cases} \theta & \text{if } j = i+1 \\ \gamma & \text{if } j = i-1 \\ 0 & \text{in all other cases.} \end{cases}$$

We leave the proof as Problem 14. The projection defined by Eq. (23.38) is illustrated by the bottom graph of Figure 23.4.

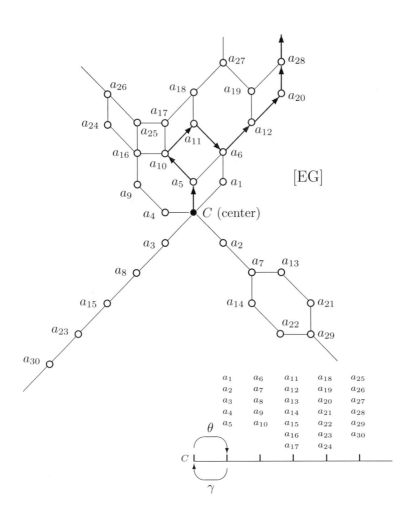

Figure 23.4: [EG] Example of a random walk on a graph satisfying the even-triangle condition. A possible realization is indicated by the thick arrows. In such a situation the graph can be projected on the nonnegative integers (see the bottom graph of the figure).

Problems

1. Rewrite Axiom [W3] of 23.1 so as to make it consistent with the situation of Example 21.4.

2. In Definition 23.1, suppose that $\theta + \gamma = 1$. Under that assumption, do all the paths from the origin to the point A in Fig. 23.1 have the same probability? Prove your answer.

3. In the style of Formula (23.4) write explicitly the expressions defining the following events: (a) the particle never hits two bounds $r, q \in \mathbb{Z}$; (b) the two bounds $r, q \in \mathbb{Z}$ are absorbing barriers, that is, the particle remains stuck to a bound as soon as it hits one; and (c) the particle always remains on one side of the origin after leaving it.

4. For the probability space defined in 23.1, define two disjoint events of measure zero.

5. Solve Problem 23.6 in Case 2 (that is, when $\theta = \gamma = \frac{1}{2}$) by applying L'Hospital's rule to Eq. (23.13).

6. Prove that (23.15) is the unique solution of Problem 23.6 in Case 2 (that is, when $\theta = \gamma = \frac{1}{2}$).

7. (a) Rephrase the solution of Problem 23.6 in the language of the coin tossing game of 21.4. (b) Suppose that Peter begins the game with $10, and gives a dollar to Paul if a head comes up, and receives a dollar otherwise. Peter adopts the strategy to stop the game if either he loses his $10 (ruin) or his purse reaches $20, that is, his initial $10 plus a gain of $10. Compute the probability of Peter's ruin. We assume that the probability of a head is $\frac{2}{3}$ and that the fortune of Paul is irrelevant (that is, essentially infinite). (c) What is the probability of Peter's ruin if he starts the game with $15? (d) With $19?

8. (a) Rephrase the solution of Problem 23.9 in the language of the coin tossing game of 21.4. (b)-(c) In the situations described in (c) and (d) in Problem 7, respectively, compute the expected times until the end of the game.

9. (a) Axiomatize the random walk on the planar grid illustrated by Figure 23.2. (b) For the random walk on a planar grid that you defined, formulate problems in the spirit of Problem 23.6 and 23.9.

10. Consider a random walk on $\mathbb{Z}$ in which $\theta = \gamma = \frac{1}{2}$. For each positive integer n, find $\mathbb{P}(\mathbf{X}_n = 0)$, that is, find the probability that the random walk is at the origin at time n (whether or not it went anywhere else between trial 0 and trial n).

11. Verify that the graph defined in Example 23.13 is connected, discrete, and satisfies the even-triangle property.

12. Verify that the function d defined in 23.12 is a distance function on the set of vertices; that is, for any vertices x, y and z, we must have

    (a) $d(x, y) > 0$ if $x \neq y$; $d(x, x) = 0$;

    (b) $d(x, y) = d(y, x)$;

    (c) $d(x, y) + d(y, z) \geq d(x, z)$.

13. In Example [EG] of Figure 23.4, what is the probability that the particle is on vertex a_{27} on trial 5?

14. Check that Theorem 23.19 is correct.

Chapter 24

Poisson Processes

The previous three chapters were devoted to stochastic models for certain phenomena evolving in 'discrete' time $0, 1, \ldots, n, \ldots$, with these labels possibly marking, for example, the number of days or minutes of observation, or the number of trials of an experiment. In many important cases, however, we are interested in phenomena evolving in situations where time flows continuously, with no natural parsing in discrete units. The focus of the chapter is on a particular class of such phenomena, where the events recorded are 'punctual' in the sense that their duration is so short that it can be ignored. Moreover, the occurrence of each event is measured by a function taking only two possible values, such as '0-1' or 'on-off.' These situations are illustrated by the first two examples below.

Examples of Continuous Time Phenomena

24.1. Example. (RADIOACTIVE DECAY.) A Geiger counter is recording the arrival of particles from a radioactive source. By convention, the arrival of each particle is regarded as a punctual event. A sequence of such punctual events, each of which is marked by the time of arrival of the particle, can be regarded as a realization of a 'continuous time stochastic process' (in the sense of Def. 24.6). Figure 24.1 displays a exemplary picture of such a realization. According to this picture, the Geiger counter has recorded the arrival of a particle at times $t_1, t_2, t_3, \ldots, t_n, \ldots$ The ordinate indicates at each time $t \geq 0$ whether the counter is 'on' (a particle has arrived), or 'off' (no arrival).

Figure 24.1: Exemplary picture of a realization of a punctual, two-valued continuous time stochastic process (see Definition 24.6). The punctual events have been observed at times t_1, t_2, t_3, ... , t_n, ...

As indicated by its name, the Geiger counter keeps a running count of the total number of events that have occurred since the beginning of the recording period. The relevant picture is in Fig. 24.2 and corresponds to that of Fig. 24.1 (that is, the two pictures contain exactly the same information).

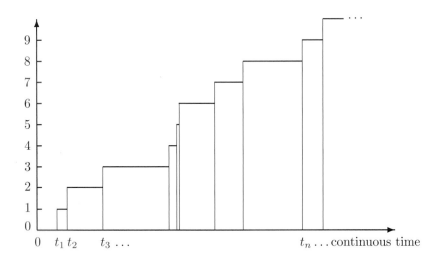

Figure 24.2: Running count of the total number of punctual events accumulated since the beginning in the picture of Fig. 24.1.

24.2. Example. (ELECTRICAL ACTIVITY OF A NERVE CELL.) Using an electrode implanted in a single neurone located in the cortex of a sedated animal, a physiologist is recording the spontaneous[1]

---
[1]That is, in the absence of any external stimulation.

electrical activity of the cell. Typically, this activity takes the form of a seemingly random firing of the cell, each firing producing a single spike in the recording. Because each spike occurs in a very short span of time, it is often assumed for theoretical purposes that it is instantaneous. In other words, such spikes are regarded as punctual events. An instance of such a recording may then be idealized by representations of the kind pictured in Figs. 24.1 and 24.2.

The next example is only slightly more complicated and illustrates a situation which is already beyond the scope of this chapter. The phenomenon under study is no longer punctual. We still analyze it in terms of a function taking only two values ('on' or 'off'), but we also measure the duration of the 'off' periods.

24.3. Example. (MACHINE FAILURE.) A controller is in charge of checking the functioning state of a piece of equipment. Part of his responsibility involves recording the exact time of any failure of the machine, and the duration of the repair. A graphic example of such a record is displayed in Fig. 24.3. It shows that the first two failures of the machine were at time t_1 and t_3. The corresponding repairs took place at times t_2 and t_4, respectively. The durations of these repairs were thus $t_2 - t_1$ and $t_4 - t_3$.

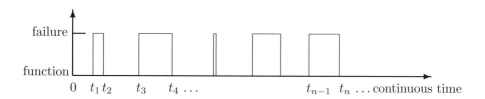

Figure 24.3: Exemplary picture of the failure and repair times of a piece of equipment. The first two failures were observed at times t_1 and t_3, and the corresponding repairs at times t_2 and t_4. The durations of the repairs were $t_2 - t_1$, $t_4 - t_3$, etc. (See Example 24.3.)

Obviously, considerably more complex situations exist in which a phenomenon is measured not only continuously, but also in terms of one or possibly several real valued functions. Even though such situations are not covered in this chapter, we give two examples for the sake of contrast.

24.4. Example. (THE ACTIVITY OF THE BRAIN.) Even during sleep, small currents can be continuously recorded at the surface of the brain by electroencephalographic (EEG) methods. The current is transmitted to the measuring device by small wires attached to electrodes fixed to the surface of the skull. To the extent that such activity displays irregularities unrelated to events detectable by an observer, such a recording can be regarded as a realization of a stochastic process.

24.5. Example. (THE STOCK MARKET.) The stock market is evaluated by various indices. The Dow Jones Industrial Average, for example, is the average (computed at the end of every trading day) of the stock prices of 30 carefully selected large companies. As many investors would confirm, the day-to-day evolution of this index is unpredictable, and we can regard a particular sequence of values of the Dow Jones as a realization of a stochastic process. From a theoretical standpoint, even though a new value is computed every day, it makes sense to consider the evolution of the index as a continuous process.

The models discussed in this chapter are appropriate for situations such as those of Examples 24.1 and 24.2, that is, the evolving phenomenon is captured by a sequence of punctual events occuring in continuous time and measured by a two-valued ('on-off') function. In addition, a pervasive condition of independence between the events lies at the core of the model. As a consequence of these assumptions, the number of events counted within any fixed interval of time has a Poisson distribution (cf. 8.12) with a parameter proportional to the length of the interval. This explains the name 'Poisson' processes given to such models. Finally, the model is formulated in terms of a collection of 'counting' random variables—indexed by the time $t \geq 0$—which keeps track of the total number of punctual events accumulated since the beginning. The next section formalizes these ideas.

On the basis of this description and of the examples, it may seem that Poisson processes are relatively trivial systems which may be too simple to be useful for important applications. Actually, as we shall see, the concept of a Poisson process is a very rich one, which provides an important stepping stone for the study of many much more complex phenomena.

Basic Concepts

24.6. Definition. A collection of random variables $(\mathbf{X}_t)_{t\in T}$ is called a *stochastic process* if either $T = \{0, 1, \ldots, n, \ldots\}$ or $T = \mathbb{R}_+$. The collection $(\mathbf{X}_s)_{s\in T}$ is referred to as a *discrete time* stochastic process if $T = \{0, 1, \ldots, n \ldots\}$ and as a *continuous time* stochastic process if $T = \mathbb{R}_+$. Thus, the Markov chains discussed in Chapters 21, 22, and 23 are instances of discrete time stochastic processes.

24.7. Definition. We say that a continuous time stochastic process $(\mathbf{X}_t)_{t\in\mathbb{R}_+}$ has *independent increments* if for any positive integer n and any choice of $n+1$ indices $t_0, t_1, \ldots, t_n$, the n random variables

$$\mathbf{X}_{t_1} - \mathbf{X}_{t_0}, \ldots, \mathbf{X}_{t_n} - \mathbf{X}_{t_{n-1}}$$

are independent. A continuous time stochastic process $(\mathbf{X}_t)_{t\in\mathbb{R}_+}$ has *stationary increments* if, for any $t \geq 0$ and $\delta > 0$, the distribution of the difference $\mathbf{X}_{t+\delta} - \mathbf{X}_t$ does not depend on t.

There are various ways of defining a Poisson process. Here is a simple one.

24.8. Definition. A continuous time stochastic process $(\mathbf{N}_t)_{t\in\mathbb{R}_+}$ is a *(homogeneous) Poisson* process with *(intensity) parameter* $\lambda > 0$ if the following three conditions hold:

[Q1] $\mathbf{N}_0 = 0$;

[Q2] the stochastic process $(\mathbf{N}_t)_{t\in\mathbb{R}_+}$ has independent increments;

[Q3] for any $s, t \in \mathbb{R}_+$ with $s < t$, the random variable $\mathbf{N}_t - \mathbf{N}_s$ has a Poisson distribution with parameter $\lambda(t - s)$, that is, for $k = 0, 1, \ldots$, we have (cf. 8.12 and Table 15.1)

$$\mathbb{P}(\mathbf{N}_t - \mathbf{N}_s = k) = e^{\lambda(t-s)} \frac{(\lambda(t - s))^k}{k!},$$

$$E(\mathbf{N}_t - \mathbf{N}_s) = Var(\mathbf{N}_t - \mathbf{N}_s) = \lambda(t - s).$$

24.9. Remark. Notice that, as a consequence of Axioms [Q2] and [Q3], the stochastic process $(\mathbf{N}_t)_{t\in\mathbb{R}_+}$ has stationary independent increments. The qualifier 'homogeneous' in Def. 24.8, which is usually omitted, refers to the fact that the density of the punctual events (measured by the intensity parameter λ) is constant over time (because the difference $\mathbf{N}_{t+\delta} - \mathbf{N}_t$ only depends upon δ). This assumption is dropped in the concept of a 'nonhomonegeous Poisson' process. Consider a continuous time stochastic process $(\mathbf{N}_t)_{t \in \mathbb{R}_+}$ satisfying Axioms [Q1] and [Q2] of Def. 24.8 together with the two axioms:

[Q3'] for any real number $t > 0$, we have $0 < \mathbb{P}(\mathbf{N}_t > 0) < 1$;

[Q4'] For any $t \in \mathbb{R}_+$,
$$\lim_{\delta \to 0} \frac{\mathbb{P}(\mathbf{N}_{t+\delta} - \mathbf{N}_t \geq 2)}{\mathbb{P}(\mathbf{N}_{t+\delta} - \mathbf{N}_t = 1)} = 0. \tag{24.1}$$

Axiom [Q3'] states that, in any open interval $]0, t[$ (no matter how small t might be), there is a positive probability less than 1 that at least one punctual event occurs. Regarding the ratio in Eq. (24.1) of Axiom [Q4'], notice that because the event $\mathbf{N}_{t+\delta} - \mathbf{N}_t = 1$ is included in the event $\mathbf{N}_{t+\delta} - \mathbf{N}_t \geq 1$, we must have

$$\frac{\mathbb{P}(\mathbf{N}_{t+\delta} - \mathbf{N}_t \geq 2)}{\mathbb{P}(\mathbf{N}_{t+\delta} - \mathbf{N}_t = 1)} \geq \frac{\mathbb{P}(\mathbf{N}_{t+\delta} - \mathbf{N}_t \geq 2)}{\mathbb{P}(\mathbf{N}_{t+\delta} - \mathbf{N}_t \geq 1)}. \tag{24.2}$$

The right-hand side of (24.2) is the conditional probability of observing at least two punctual events in an interval $]t, t + \delta[$, given that at least one such punctual event is observed. According to Axiom [Q4'], the left-hand side of (24.2) vanishes as $\delta \to 0$; thus, so does the right-hand side. This remark supports the interpretation of Axiom [Q4'] as asserting that when the interval $]t, t + \delta[$ becomes vanishingly small, at most one punctual event can arise.[2] Axioms [Q1], [Q2], [Q3'], and [Q4'] are consistent with a process in which the punctual events may not occur with equal density over time.

Remarkably, an homogeneous Poisson process is obtained when we add to Axioms [Q1], [Q2], [Q3'], and [Q4'] the assumption that $(\mathbf{N}_t)_{t \in \mathbb{R}_+}$ has stationary increments in the sense of Def. 24.7. (For a proof of this fact, see for example, Parzen, 1994b). A *nonhomogeneous Poisson process* is a continuous time stochastic process $(\mathbf{N}_t)_{t \in \mathbb{R}_+}$ which satisfies Axioms [Q1], [Q2], [Q3'], and [Q4'] and does not have stationary increments. In this case, for $s < t$, the difference $\mathbf{N}_t - \mathbf{N}_s$ is still Poisson distributed as in [Q3] but with a mean $\int_s^t \lambda(v) \, d(v)$ where λ is the *arrival time function*.

---

[2]The reader may remember a similar assumption made in 8.12 to derive the Poisson distribution as a limiting case of a binomial distribution.

Summing Independent Poisson Processes

In some situations, we are led to ask whether the observed punctual events could have been produced by the aggregation of several distinct random mechanisms. In the radioactive decay of Example 24.1, for instance, one might imagine that several unrelated radioactive sources are generating the particles hitting the Geiger counter. Even if each of the sources induces a Poisson process, the parameters λ of these Poisson processes may be different. Nevertheless, if the Poisson processes are independent in the sense of Def. 24.10 below, then the sum of the Poisson processes is itself a Poisson process, with an intensity parameter which is the sum of the intensity parameters of the component processes. This result is expressed precisely in Def. 24.10 and Theorem 24.12.

24.10. Definition. Let $(\mathbf{X}_t)_{t \in \mathbb{R}_+}$ and $(\mathbf{Y}_t)_{t \in \mathbb{R}_+}$ be two continuous time stochastic processes. We say that these processes are *independent* if for any choice of $s_1, \dots, s_n \in \mathbb{R}_+$ and $t_1, \dots, t_m \in \mathbb{R}_+$, and of continuous functions $G : \mathbb{R}^n \to \mathbb{R}$ and $H : \mathbb{R}^m \to \mathbb{R}$, the random variables $G(\mathbf{X}_{s_1}, \dots, \mathbf{X}_{s_n})$ and $H(\mathbf{Y}_{t_1}, \dots, \mathbf{Y}_{t_m})$ are independent.

24.11. Remark. This definition assumes implicitly that the two processes $(\mathbf{X}_t)_{t \in \mathbb{R}_+}$ and $(\mathbf{Y}_t)_{t \in \mathbb{R}_+}$ are *jointly distributed* in the sense that for any choice of numbers

$$s_1, \dots, s_n \in \mathbb{R}_+, \qquad s_1', \dots, s_n' \in \mathbb{R},$$
$$t_1, \dots, t_m \in \mathbb{R}_+, \qquad t_1', \dots, t_m' \in \mathbb{R},$$

the following joint probability is defined

$$\mathbb{P}(\mathbf{X}_{s_1} \le s_1', \dots, \mathbf{X}_{s_n} \le s_n', \mathbf{Y}_{t_1} \le t_1', \dots, \mathbf{Y}_{t_m} \le t_m').$$

24.12. Theorem. *If* $(\mathbf{N}_{1,t})_{t \in \mathbb{R}_+}$ *and* $(\mathbf{N}_{2,t})_{t \in \mathbb{R}_+}$ *are two independent Poisson processes, with respective intensity parameters* λ_1 *and* λ_2, *then* $(\mathbf{N}_{1,t} + \mathbf{N}_{2,t})_{t \in \mathbb{R}_+}$ *is also a Poisson process, with intensity parameter* $\lambda_1 + \lambda_2$.

Note that the sum $\mathbf{N}_{1,t} + \mathbf{N}_{2,t}$ is sometimes called a *superposition* of the two independent Poisson processes $(\mathbf{N}_{1,t})_{t \in \mathbb{R}_+}$ and $(\mathbf{N}_{2,t})_{t \in \mathbb{R}_+}$.

PROOF. We must show that Axioms [Q1]-[Q3] of Def. 24.8 hold for the continuous time stochastic process $(\mathbf{N}_{1,t} + \mathbf{N}_{2,t})_{t \in \mathbb{R}_+}$. Axiom [Q1] is immediate: since both $(\mathbf{N}_{1,t})_{t \in \mathbb{R}_+}$ and $(\mathbf{N}_{2,t})_{t \in \mathbb{R}_+}$ satisfy

[Q1], we get $\mathbf{N}_{1,0} + \mathbf{N}_{2,0} = 0 + 0 = 0$. Because the two processes $(\mathbf{N}_{1,t})_{t\in\mathbb{R}_+}$ and $(\mathbf{N}_{2,t})_{t\in\mathbb{R}_+}$ are independent, for any $t \in \mathbb{R}_+$ the random variable $\mathbf{N}_{1,t} + \mathbf{N}_{2,t}$ is the sum of two independent Poisson random variables with respective parameters λ_1 and λ_2. Using Example 18.7 and Theorem 18.8, we can assert that $\mathbf{N}_{1,t} + \mathbf{N}_{2,t}$ is also a Poisson random variable, with parameter $\lambda_1 + \lambda_2$. We conclude that $(\mathbf{N}_{1,t} + \mathbf{N}_{2,t})_{t\in\mathbb{R}_+}$ also satisfies [Q3]. It remains to show that $(\mathbf{N}_{1,t} + \mathbf{N}_{2,t})_{t\in\mathbb{R}_+}$ has independent increments. We leave this task to the reader (see Problem 1).   □

Random Selection of Punctual Events

This is sometimes referred to as a random 'thinning' of a Poisson process. Consider a situation in which only some of the punctual events of a Poisson process $(\mathbf{N}_t)_{t\in\mathbb{R}_+}$ are detected by a recording mechanism. Specifically, suppose that there is a probability $p < 1$ that a punctual event is detected, this probability being constant for all punctual events. Moreover, the detection of any punctual event of the process $(\mathbf{N}_t)_{t\in\mathbb{R}_+}$ is independent of the detection of any other event. (This might be a plausible model for the radioactive decay of Example 24.1, for instance, in a case where a defective Geiger counter has been used that would only detect the arrival of a particle with a probability equal to p.) Let us denote by $(\mathbf{M}_t)_{t\in\mathbb{R}_+}$ the stochastic process resulting from the detection. If k punctual events have occurred in the interval of time $]t, t+\delta[$, then the number of detected events is binomially distributed with parameters k and p. Formally, we have

$$\mathbb{P}(\mathbf{M}_{t+\delta} - \mathbf{M}_t = j \mid \mathbf{N}_{t+\delta} - \mathbf{N}_t = k) = \binom{k}{j} p^j (1-p)^{k-j}. \quad (24.3)$$

Under these hypotheses and assuming that the Poisson process $(\mathbf{N}_t)_{t\in\mathbb{R}_+}$ has intensity λ, the continuous time stochastic process $(\mathbf{M}_t)_{t\in\mathbb{R}_+}$ is also a Poisson process which has parameter $p\lambda$. Indeed, it is clear that $\mathbf{M}_0 = 0$, and it is easily verified that $(\mathbf{M}_t)_{t\in\mathbb{R}_+}$ has independent increments. Moreover, since

$$\mathbb{P}(\mathbf{N}_{t+\delta} - \mathbf{N}_t = k) = e^{-\delta\lambda}\frac{(\lambda\delta)^k}{k!},$$

the Theorem of Total Probabilities (cf. Theorem 9.8 and Remark 9.9) together with Eq. (24.3) give successively

$$\mathbb{P}(\mathbf{M}_{t+\delta} - \mathbf{M}_t = j)$$

$$= \sum_{k=j}^{\infty} \mathbb{P}(\mathbf{M}_{t+\delta} - \mathbf{M}_t = j \mid \mathbf{N}_{t+\delta} - \mathbf{N}_t = k)\mathbb{P}(\mathbf{N}_{t+\delta} - \mathbf{N}_t = k)$$

$$= \sum_{k=j}^{\infty} \binom{k}{j} p^j (1-p)^{k-j} e^{-\delta\lambda} \frac{(\lambda\delta)^k}{k!}$$

$$= e^{-\delta\lambda} \frac{(p\lambda\delta)^j}{j!} \sum_{k=j}^{\infty} \frac{(\delta\lambda(1-p))^{k-j}}{(k-j)!} \tag{24.4}$$

$$= e^{-\delta\lambda} \frac{(p\lambda\delta)^j}{j!} e^{(1-p)\delta\lambda} \tag{24.5}$$

$$= e^{-\delta p\lambda} \frac{(p\lambda\delta)^j}{j!}. \tag{24.6}$$

We conclude that the continuous time stochastic process $(\mathbf{M}_t)_{t\in\mathbb{R}_+}$ resulting from the random selection—with probability p—of punctual events from the Poisson process $(\mathbf{N}_t)_{t\in\mathbb{R}_+}$ is itself a Poisson process with intensity parameter $p\lambda$. In Problem 2, the student is required to justify some of the steps in the above argument.

Interarrival Times

A change of outlook is taking place here: Rather than focusing on the number of punctual events observed in an interval of time, which has a Poisson distribution according to Axiom [Q3] in Def. 24.8, we consider instead the time elapsing beween punctual events. It turns out that, in a homogeneous Poisson process, the time between two consecutive punctual events has an exponential distribution with parameter $\lambda > 0$, where λ is the intensity parameter of the Poisson process. This property is a fundamental aspect of Poisson processes. In fact, we can start from the assumption that the interarrival times of consecutive punctual events are independent exponentials with the same parameter and derive a Poisson process from this premise. We do so in this section.

Let $\mathbf{T}_1, \ldots, \mathbf{T}_k, \ldots$ be a sequence of identically distributed, independent random variables. We suppose that their common distribution is exponential with parameter $\lambda > 0$ (cf. 14.6), that is, for $k = 1, 2, \ldots,$

$$\mathbb{P}(\mathbf{T}_k \leq t) = \begin{cases} 1 - e^{-\lambda t} & \text{if } t \geq 0; \\ 0 & \text{otherwise.} \end{cases} \tag{24.7}$$

We have thus $\mathbb{P}(\mathbf{T}_k > t) = e^{-\lambda t}$ (for $t \geq 0$), and because the random variables $\mathbf{T}_k$ are independent,

$$\mathbb{P}(\mathbf{T}_1 > t_1, \ldots, \mathbf{T}_k > t_k) = e^{-\lambda(t_1 + \ldots + t_k)}. \tag{24.8}$$

Setting

$$\mathbf{S}_0 = 0, \tag{24.9}$$
$$\mathbf{S}_k = \mathbf{T}_1 + \ldots + \mathbf{T}_k \qquad (k \in \mathbb{N}), \tag{24.10}$$

the sum $\mathbf{S}_k$ measures the time required for the occurrence of the kth punctual event, counted from the time $t = 0$. We now define, for all $t \in \mathbb{R}_+$ the random variable $\mathbf{N}_t$ with range $\mathbb{N}_0$ by the equivalence

$$\mathbf{N}_t = k \quad \Longleftrightarrow \quad \mathbf{S}_k \leq t < \mathbf{S}_{k+1} \qquad (k \in \mathbb{N}_0). \tag{24.11}$$

The two formulas '$\mathbf{N}_t = k$' and '$\mathbf{S}_k \leq t < \mathbf{S}_{k+1}$' denote exactly the same event in the common probability space of all the random variables $\mathbf{N}_t$ and $\mathbf{S}_k$. Accordingly, we obtain

$$\mathbb{P}(\mathbf{N}_t = k) = \mathbb{P}(\mathbf{S}_k \leq t < \mathbf{S}_{k+1}). \tag{24.12}$$

In principle, we can thus derive the probability distribution of the random variable $\mathbf{N}_t$ from the probability distribution of the random variables $\mathbf{S}_k$ and $\mathbf{S}_{k+1}$. We begin by finding the probability distribution of a random variable $\mathbf{S}_k$ which is the sum of k independent exponential random variables with the same parameter λ (cf. (24.7) and (24.10)). The following definition and theorem will be instrumental for this purpose.

24.13. Definition. A random variable $\mathbf{G} = \mathbf{G}(k, \lambda)$ has a *gamma distribution* with parameters $k \in \mathbb{N}$ and $\lambda > 0$ if its density has the form

$$f_{\mathbf{G}}(t) = \begin{cases} \frac{\lambda^k t^{k-1}}{(k-1)!} e^{-\lambda t} & \text{if } t \in \mathbb{R}_+, \\ 0 & \text{otherwise.} \end{cases} \tag{24.13}$$

24.14. Remarks. (a) For $k = 1$, Eq. (24.13) reduces to $\lambda e^{-\lambda t}$ for $t \in \mathbb{R}_+$. Thus, the exponential random variable with parameter $\lambda > 0$ is a special case of the gamma random variable $\mathbf{G}(k, \lambda)$ where $k = 1$.

(b) A more general version of the gamma distribution exists in which the integer k in (24.13) is replaced by a real number $s > 0$ and the factorial $(k - 1)!$ by the gamma function

$$\Gamma(s) = \int_0^\infty x^{s-1} e^{-x} dx \qquad (s > 0)$$

(see, for example, Parzen, 1994a).

24.15. Theorem. *The moment generating function of the gamma random variable* $\mathbf{G} = \mathbf{G}(k, \lambda)$ *with a density defined by (24.13), is defined by the equation*

$$M_{\mathbf{G}}(\theta) = \left(\frac{\lambda}{\lambda - \theta}\right)^k.$$

PROOF. By the definition of the M.G.F. in 18.1, we must prove that

$$M_{\mathbf{G}}(\theta) = E(e^{\theta \mathbf{G}}) = \int_0^\infty \frac{\lambda^k t^{k-1}}{(k-1)!} e^{\theta t} e^{-\lambda t} dt = \left(\frac{\lambda}{\lambda - \theta}\right)^k. \qquad (24.14)$$

We use a detour. A basic result of calculus is that, for any $x > 0$, we have

$$\int_0^\infty e^{-xt} dt = \frac{1}{x}.$$

Taking successive derivatives in the variable x on both sides, we obtain for the $(k-1)^{\text{th}}$ derivative

$$\int_0^\infty (-t)^{k-1} e^{-xt} dt = \frac{(-1)^{k-1}(k-1)!}{x^k},$$

which implies

$$\int_0^\infty \frac{t^{k-1}}{(k-1)!} e^{-xt} dt = \frac{1}{x^k}.$$

Setting $x = \lambda - \theta$ and multiplying by λ^k on both sides yields the last equality in (24.14) and establishes the result. $\qquad \square$

24.16. Theorem. *The sum* $\mathbf{S}_k = \mathbf{T}_1 + \ldots + \mathbf{T}_k$ *of* k *independent exponential random variables with the same parameter* $\lambda > 0$ *is a gamma random variable with parameters* k *and* λ.

PROOF. From Table 15.1, we know that the M.G.F. of an exponential random $\mathbf{T}$ variable with parameter λ has the form

$$M_{\mathbf{T}}(\theta) = \frac{\lambda}{\lambda - \theta}.$$

By Theorem 18.6, we also know that the moment generating function of a sum $\mathbf{S}_k = \mathbf{T}_1 + \ldots + \mathbf{T}_k$ of k independent random variables is the product of the M.G.F.'s of these random variables, and so, if these random variables are exponential with the same parameter λ, we must have

$$M_{\mathbf{S}_k}(\theta) = \left(\frac{\lambda}{\lambda - \theta}\right)^k.$$

Note that, by Theorem 24.15, the right-hand side of this equation is the M.G.F. of a gamma random variable with parameters k and λ. Since, by Theorem 18.8, the moment generating function of a random variable uniquely characterizes this random variable, we conclude that the Theorem holds and that the sum $\mathbf{S}_k$ is indeed a gamma random variable with parameters k and λ. $\qquad\square$

Recalling Eq. (24.12) and applying Theorem 24.16, we obtain

$$
\begin{aligned}
\mathbb{P}(\mathbf{N}_t = k) &= \mathbb{P}(\mathbf{S}_k \leq t < \mathbf{S}_{k+1}) \\
&= \mathbb{P}(\mathbf{S}_k \leq t) - \mathbb{P}(\mathbf{S}_{k+1} \leq t) \\
&= \int_0^t \frac{\lambda^k s^{k-1}}{(k-1)!} e^{-\lambda s}\, ds - \int_0^t \frac{\lambda^{k+1} s^k}{k!} e^{-\lambda s}\, ds. \quad (24.15)
\end{aligned}
$$

Integrating the first integral by parts (see Problem 3) gives

$$\int_0^t \frac{\lambda^k s^{k-1}}{(k-1)!} e^{-\lambda s}\, ds = e^{-\lambda t} \frac{(t\lambda)^k}{k!} + \int_0^t \frac{\lambda^{k+1} s^k}{k!} e^{-\lambda s}\, ds, \quad (24.16)$$

which implies, putting (24.15) and (24.16) together,

$$\mathbb{P}(\mathbf{N}_t = k) = e^{-\lambda t} \frac{(t\lambda)^k}{k!}. \quad (24.17)$$

We summarize and go one step further.

24.17. Theorem. *Let $(\mathbf{T}_k)_{k \in \mathbb{N}}$ be a sequence of independent exponential random variables with the same parameter λ. Define*

$$\mathbf{S}_0 = 0, \quad \mathbf{S}_k = \mathbf{T}_1 + \ldots + \mathbf{T}_k \qquad (k \in \mathbb{N}) \qquad (24.18)$$
$$\mathbf{N}_t = k \iff \mathbf{S}_k \leq t < \mathbf{S}_{k+1} \qquad (t \in \mathbb{R}_+). \qquad (24.19)$$

Then,

(i) *for any $k \in \mathbb{N}$, the random variable $\mathbf{S}_k$ is gamma distributed;*

(ii) *for any $t \in \mathbb{R}_+$, the random variable $\mathbf{N}_t$ is Poisson distributed with parameter λt, that is, Eq. (24.17) holds;*

(iii) *in fact, the continuous time stochastic process $(\mathbf{N}_t)_{t \in \mathbb{R}_+}$ is a Poisson process.*

SKETCH OF PROOF. Statements (i) and (ii) have been established earlier: The first is just a restatement of Theorem 24.16, and for the second, see Eq. (24.17). To show that $(\mathbf{N}_t)_{t \in \mathbb{R}_+}$ is a Poisson process, we may verify that Axioms [Q1], [Q2], and [Q3] of Def. 24.8 hold. Axiom [Q1] is immediate from the definition of $\mathbf{N}_0$: The right-hand side of the equivalence (24.19) holds for $t = 0$ only when $\mathbf{S}_0 = 0 \leq \mathbf{S}_1 = \mathbf{T}_1$. Using the fact that the random variables $\mathbf{T}_k$ are independent exponentials with the same parameter λ, we can derive not only that $(\mathbf{N}_t)_{t \in \mathbb{R}_+}$ has independent increments (and thus, [Q2] holds), but also that these increments are stationary. In turn, this implies that, for $s < t$, the distribution of the difference $\mathbf{N}_t - \mathbf{N}_s$ is the same as that of $\mathbf{N}_{t-s} - \mathbf{N}_0$, which we know, by our earlier argument, to be Poisson with parameter $\lambda(t - s)$ (see Eq. (24.17)). Thus, [Q3] holds, and (iii) is also true. □

24.18. Remark. There is a partial converse of Theorem 24.17. From the assumptions defining a Poisson process with parameter $\lambda > 0$, it can be derived that the intervals between any two successive punctual events are independent exponential random variables with the same parameter λ and also that the time between the kth punctual event and the $(k + j)$th punctual events is a gamma random variable with parameters j and λ. We do not include proofs of these facts here.

In view of the complexity of the Poisson process and its multiple facets, we provide in the last section of this chapter a summary of the main facts, whether or not they have been proven here.

Uniform Distribution of Arrival Times

In a Poisson process, if we only know that k punctual events are realized in a certain interval of time, exactly when in that interval these punctual events have occurred is maximally uncertain, and therefore, the distribution of these times of occurrence should, intuitively, conform to a sampling of k points in a uniform distribution with parameters s and t. In the theorem below, we deal with the case $k = 1$ and leave the more general case as Problem 4. The situation of the theorem is illustrated by Fig. 24.4.

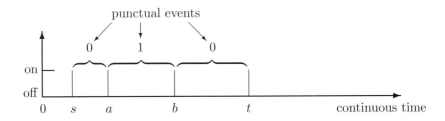

Figure 24.4: **Graph of the situation in which exactly one punctual event is realized in the interval of time** $]s, t[$, **and one asks for the probability that this event occurs in the subinterval** $]a, b[$.

24.19. Theorem. *Suppose that* $(\mathbf{N}_t)_{t \in \mathbb{R}_+}$ *is a Poisson process, and let* $0 \leq s \leq a < b \leq t$ *be real numbers. Given that exactly one punctual event occurs in the interval of time* $]s, t[$, *the conditional probability that this event occurs in the subinterval* $]a, b[$ *is* $\frac{b-a}{t-s}$.

Thus, when $a = s$, the conditional probability of the theorem is the uniform distribution function with parameters s and t, evaluated at the point b.

PROOF. The following four events enter in the formulation of the theorem:

A	0 EVENTS IN THE INTERVAL $]s, a[$
B	1 EVENT IN THE INTERVAL $]a, b[$
C	0 EVENTS IN THE INTERVAL $]b, t[$
D	1 EVENT IN THE INTERVAL $]s, t[$.

We must prove that

$$\mathbb{P}(A \cap B \cap C \mid D) = \frac{b - a}{t - s}.$$

Notice that $A \cap B \cap C \subseteq D$, and that A, B, and C are independent events. By Axiom [Q3], the probability that exactly k events occur in some interval of length δ is equal to $e^{-\delta\lambda} \frac{(\delta\lambda)^k}{k!}$. We obtain

$$
\begin{aligned}
\mathbb{P}(A \cap B \cap C \mid D) &= \frac{\mathbb{P}(A \cap B \cap C \cap D)}{\mathbb{P}(D)} \\
&= \frac{\mathbb{P}(A)\mathbb{P}(B)\mathbb{P}(C)}{\mathbb{P}(D)} \\
&= \frac{e^{-(a-s)\lambda} e^{-(b-a)\lambda}(b - a)\lambda e^{-(t-b)\lambda}}{e^{-(t-s)\lambda}(t - s)\lambda} \\
&= \frac{b - a}{t - s},
\end{aligned}
$$

as asserted. $\qquad\qquad\qquad\qquad\qquad\qquad\qquad\qquad\qquad\qquad\qquad$ $\square$

Summary of the Poisson Process

Let $(\mathbf{N}_t)_{t \in \mathbb{R}_+}$ be a Poisson process with intensity parameter $\lambda > 0$, as defined by Axioms [Q1], [Q2], and [Q3] of Def. 24.8. The following facts follow:

1. For any $k \in \mathbb{N}$, the interval of time $\mathbf{T}_k$ between the $(k-1)$th and the kth punctual events, or (if $k = 1$) between the beginning and the first punctual event, is an exponential random variable with parameter λ. The random variables $\mathbf{T}_k$ ($k \in \mathbb{N}$) are independent. (See 14.6 and 24.18.)

2. For $j, k \in \mathbb{N}$, the interval of time $\mathbf{G}_k(j)$ between the $(k-1)^{\text{th}}$ and the $(k - 1 + j)^{\text{th}}$ punctual events, or (if $k = 1$) between the beginning and the jth punctual event, is a gamma random variable with parameters λ and j. For appropriate values of pairs (k, j) (cf. Problem 5), the random variables $\mathbf{G}_k(j)$ are also independent. (See 24.13 and 24.18.)[3]

---

[3]Fact 2 subsumes of course Fact 1 (take $j = 1$).

3. For any arbitrarily chosen time t, the time elapsing between t and the jth punctual event counted after time t is also gamma distributed, with the same parameters λ and j. (This results from the memory-less property of the exponential distribution; see 14.7.)

4. Suppose that we only retain some of the punctual events generated by the process $(\mathbf{N}_t)_{t\in\mathbb{R}_+}$, the selection being performed with a constant probability p and each selection being independent of any other selection. Then, such a random selection (also called 'thinning') induces a Poisson process with parameter $p\lambda$. (See Eqs. (24.3) and (24.6).)

5. The sum (or superposition) of two independent Poisson processes with respective parameters λ_1 and λ_2 is also a Poisson process, which has parameter $\lambda_1 + \lambda_2$. (See 24.10, 24.12.)

6. The arrival times of the punctual events are uniformly distributed in the following sense: Given that k events occur in some interval of time $]s, t[$, the exact arrival times of these k events in that interval are governed by a random sampling in a uniform distribution with parameters s and t. (See Theorem 24.19 and Problem 4.)

Problems

1. Complete the proof of Theorem 24.12 and show that the continuous time stochastic process $(\mathbf{N}_{1,t} + \mathbf{N}_{2,t})_{t\in\mathbb{R}_+}$ has independent increments.

2. Verify the steps leading to Eq. (24.6) describing the result of the random selection of punctual events in a Poisson process. Specifically, what is the argument used to derive (24.5) from (24.4)?

3. Use integration by parts to show that $\int_0^t \frac{\lambda^k s^{k-1}}{(k-1)!} e^{-\lambda s}\, ds = e^{-\lambda t} \frac{(t\lambda)^k}{k!}$.

4. Generalizing the argument used in the proof of Theorem 24.19, compute the probability that, given that exactly k events occur in the interval $]s, t[$, these event occur in the subinterval $]a, b[$, with $0 \le s \le a < b \le t$.

5. Formulate an appropriate condition of independence of the gamma random variables $\mathbf{G}_k(j)$ in Fact 2 of the last section of the chapter.

6. Find the expectation and the variance of a gamma random variable with parameters $k \in \mathbb{N}$ and $\lambda > 0$.

7. Suppose that the arrival of particles at a Geiger counter is a Poisson process with intensity λ arrivals per second and that the counter actually detects the arrival of a particle with probability p. (a) What is the probability that no particles arrive in the first minute? (b) What is the probability that no particles are detected in the first minute? (c) What is the probability that no particles are detected between the ninth and tenth minutes? (d) What is the expected number of particles detected in the first s seconds? (e) What is the variance of this number?

8. Suppose that misstatements in a speech given by a politician occur as a Poisson process with parameter 0.1 errors per minute. (a) What is the expected time (from the beginning of the speech) of the first error? Given that there were exactly two errors in the first minute, find: (b) the probability that there were exactly two errors in the first 15 seconds; (c) the probability that there was at least one error in the first 15 seconds; (d) the expected time (from the beginning of the speech) of the third error.

9. Suppose that insects are trapped in a spider's web according to a Poisson process with intensity 5 insects per week. If a curious onlooker observes the web for 24 hours, what is the probability that he will see an insect trapped during his watch?

Solutions and Hints for Selected Problems

Chapter 1

2. $\{1, 7\}$.

3. (b) Successively, we have: $A \backslash B = A \cap \overline{B} = \overline{(\overline{A} \cup \overline{\overline{B}})} = \overline{(\overline{A} \cup B)} = \overline{(B \cup \overline{A})}$. The first equality holds by definition of the set difference operation, and the second one by virtue of Equation (3.3) (see Problem 3a). The last two inequalities are obvious.

4. The Axiom of Extensionality is logically equivalent to the statement 'Two sets are distinct if at least one of them contains an element not contained in the other one.' Thus, two distinct sets cannot be both empty.

5. (c) If $A \subseteq B$, then $A \cap \overline{B} = \emptyset$ (since any x in A is also in B). Because the empty set is a subset of any set, we have $A \cap \overline{B} = \emptyset \subseteq C \cap \overline{C}$ for any set C. Conversely, if $A \cap \overline{B} \subseteq C \cap \overline{C} = \emptyset$, then $A \cap \overline{B}$ is also empty. Thus, any x in A is also in B and so $A \subseteq B$.

6. No. We have $X \times Y = Y \times X = \emptyset$ if just one of X and Y is empty. The Condition C states that both X and Y are empty. You still have to prove the statement.

7. (d) $2^{\emptyset} = \{\emptyset\}$; (e) $2^{\{\emptyset\}} = \{\emptyset, \{\emptyset\}\}$.

12. (b) The equation holds for $n = 1$ since $1 \cdot 2 = \frac{1 \cdot 2 \cdot 3}{3} = 2$. Suppose that it also holds for $n = k$, that is, $1 \cdot 2 + 2 \cdot 3 + \ldots + k(k+1) = \frac{k(k+1)(k+2)}{3}$, then

$$1 \cdot 2 + 2 \cdot 3 + \ldots + k(k+1) + (k+1)(k+2)$$
$$= \frac{k(k+1)(k+2)}{3} + (k+1)(k+2) = \frac{(k+1)(k+2)(k+3)}{3}.$$

Chapter 2

2. The sample space has 36 points. Labeling the doors A, B, C, and D, one possible outcome is $\underline{A}^* BC^\dagger D$ (using the conventions of Example 2.7).

3. (b) Let S be the set of eight swimmers in the final. If we are only interested in the top three finishers, a possible sample space is the set of all ordered triples of distinct points of S, that is, the set $\{(a,b,c)\,|\,a,b,c \in S \text{ and } a,b,c \text{ distinct}\}$. This sample space has 336 points. (This is an application of Theorem 6.14.)

   (d) There are several possibilities. If only the ratio $length : width$ is recorded, then $\mathbb{R}$ or $\mathbb{R}_+ = \,]0,\infty[$ is appropriate. If the length and width measurements are purposely retained, then $\mathbb{R}^2$ or $\mathbb{R}_+^2$ should be chosen.

5. Suppose that the rod is horizontal. Let x denote the distance from the left end of the rod to the leftmost break, and let y denote the distance from the left end of the rod to the rightmost break. The sample space is then $\{(x,y)\,|\,0 < x < y < 1\}$. (What does this space look like geometrically?) It has uncountably many points.

6. (a) An appropriate sample space is 2^C.

   (c) This event is the set $\{S \subset 2^C\,|\,c_1 \notin S\}$.

Chapter 3

1. We can take
$$\mathcal{F} = \{\emptyset, \Omega, \{a\}, \{b,c\}, \{b,c,d,e\}, \{a,d,e\}, \{d,e\}, \{a,b,c\}\}.$$
Note that $\{a,b\} \notin \mathcal{F}$; so $\mathcal{F} \neq 2^\Omega$, as required. In fact, $\mathcal{F}$ is the 'smallest' field on Ω containing the sets $\{a\}$ and $\{b,c\}$ (cf. Problem 4).

2. (b) For any $(x,y) \in \mathbb{R}^2$, the set $\{(x,y)\}$ is the complement of an open set in $\mathbb{R}^2$, and so it is a Borel set because the Borel field of $\mathbb{R}^2$ is closed under complementation.

   (c) By the argument of the preceding problem, any set $\{(x,y)\}$ with x,y in $\mathbb{Q}$ (with $\mathbb{Q}$ the set of rational numbers) is a Borel set. As $\mathbb{Q}$ is countable, so is $\mathbb{Q} \times \mathbb{Q} \subseteq \mathbb{R}^2$. (Why?) Because the Borel field of $\mathbb{R}^2$ is closed under countable union, $\mathbb{Q} \times \mathbb{Q}$ must be a Borel subset of $\mathbb{R}^2$.

3. Any σ-field in $\mathcal{G}$ contains both $\emptyset$ and Ω; thus, we also have $\emptyset, \Omega \in \cap\mathcal{G}$. Take any $A \in \cap\mathcal{G}$. For any $\mathcal{F} \in \mathcal{G}$, we have $A \in \mathcal{F}$, and so $\overline{A} \in \mathcal{F}$ because $\mathcal{F}$ is a field and as such closed under complementation. This implies $\overline{A} \in \cap\mathcal{G}$. Finally, suppose that $A_1, A_2, \ldots, A_n, \ldots$ is a countable sequence of sets all belonging to $\cap\mathcal{G}$. As each of these sets A_n, $n \in \mathbb{N}$, belongs to any of the σ-field $\mathcal{F} \in \mathcal{G}$, so does their (countable) union, that is $\cup_{n=1}^\infty A_n \in \mathcal{F}$. We conclude that $\cup_{n=1}^\infty A_n \in \cap\mathcal{G}$. Thus, $\cap\mathcal{G}$ is closed under complementation and under countable union, that is, $\cap\mathcal{G}$ is a σ-field.

4. There certainly exists some σ-field of subsets of $\mathbb{R}$ containing all the open subsets of $\mathbb{R}$, namely $2^{\mathbb{R}}$. Thus, the set $\mathcal{R}$ of ALL σ-fields on $\mathbb{R}$ containing the open sets of $\mathbb{R}$ exists. By the preceeding solution of Problem 3, we know now that $\cap \mathcal{R}$ is a σ-field which by definition is included in any σ-field of $\mathbb{R}$ and so can be called the *smallest* σ-field containing all the open sets of $\mathbb{R}$. The same argument applies to $\mathbb{R}^n$ for any $n \in \mathbb{N}$.

6. The only possible difficulty resides in the coding of the problem. We rank the three alternatives in the order 1, 2, and 3. Thus the ranking $(2, 1, 3)$ means that a_1 is ranked second, a_2 first, and a_3 third. Ties are allowed. We denote by $(2, 1, 1)$ the case where a_3 and a_2 are tied and precede a_1. The sample space is

$$\{(1, 2, 3), (1, 3, 2), (2, 1, 3), (2, 3, 1), (3, 1, 2), (3, 2, 1), (1, 1, 2),$$
$$(1, 2, 1), (2, 1, 1), (1, 2, 2), (2, 1, 2), (2, 2, 1), (1, 1, 1)\}.$$

In (d), the event 'a_2 and a_3 are tied' is denoted by

$$\{(2, 1, 1), (1, 1, 1), (1, 2, 2)\}.$$

The rest of the problem is left to the student.

Chapter 4

2. $p(a) = .6$, $p(b) = .1$, $p(c) = .3$. This is the only possibility.

4. (a) $\frac{1}{12}$; (b) $\frac{7}{12}$; (c) $\frac{3}{4}$.

7. One of the many possibilities is $p(n) = \frac{1}{2^n}$ for all $n \in \mathbb{N}$. (It is a well-known result from calculus that $\sum_{n=1}^{\infty} \frac{1}{2^n} = 1$.)

8. $\frac{1}{4}$. A correct sample space is $\{(x, y) \in \mathbb{R}^2 \mid 0 < x < y < 1\}$, with x as the leftmost break point and y as the rightmost break point. This is a right triangular, open region of $\mathbb{R}^2$ bounded by the lines $x = 0$, $y = 1$ and $x = y$. The three fragments have lengths x, $y - x$, and $1 - y$. A triangle may be formed from these fragments precisely when:
(1) $y - x < x + 1 - y$, i.e., $y < x + \frac{1}{2}$; (2) $x < y - x + 1 - y$, i.e., $x < \frac{1}{2}$; and (3) $1 - y < x + y - x$, i.e., $\frac{1}{2} < y$. This defines a right triangular subregion T of the sample space. From elementary geometry, it is easily seen that this triangle T is one of four nonoverlapping right triangles of equal area which together make the sample space. The area of T is thus $\frac{1}{4}$.

9. (a) $\frac{5}{13}$; (b) $\frac{6}{13}$; (c) $\frac{3}{13}$.

12. (a) $\frac{18}{109}$; (b) $\frac{2}{109}$; (c) $\frac{105}{109}$.

Chapter 5

1. (a) x; (b) 1; (d) $1 - x - y$.

2. first equality: by associativity of $\cup$
   second equality: by Theorem 5.1 (v)
   third equality: by Theorem 5.1 (v) and the distributivity of $\cap$ over $\cup$
   fourth equality: by Theorem 5.1 (v)
   final equality: by elementary set theory.

4. $\frac{1}{2}$.

5. (a) .32; (b) .46 (c) .85.

6. Denote by C_i the event 'candidate c_i is approved of' for $i = 1, 2, 3$. From Theorem 5.1 (v) and $0 \leq \mathbb{P} \leq 1$, we deduce that $.37 \leq \mathbb{P}(C_2 \cap C_3) \leq .60$ and $.30 \leq \mathbb{P}(C_1 \cap C_3) \leq .53$. Using Theorem 5.1 (vi), we then have

$$.32 + \mathbb{P}(C_1 \cap C_2 \cap C_3) \leq \mathbb{P}(C_1 \cup C_2 \cup C_3) \leq .76 + \mathbb{P}(C_1 \cap C_2 \cap C_3),$$

and the first inequality implies $0 \leq \mathbb{P}(C_1 \cap C_2 \cap C_3) \leq .68$.

Chapter 6

2. We only solve the first part: $11! = 39,916,800$;
   $\sqrt{2\pi} \cdot 11^{11+\frac{1}{2}} \cdot e^{-11} \approx 39,615,625$.

3. $\frac{30!}{(30-5)!} = 30 \times 29 \times 28 \times 27 \times 26 = 17,100,720$

   $\frac{30!}{(30-5)!} \approx \frac{\sqrt{2\pi} \cdot 30^{30+\frac{1}{2}} \cdot e^{-30}}{\sqrt{2\pi} \cdot 25^{25+\frac{1}{2}} \cdot e^{-25}} = \left(\frac{30}{25}\right)^{25+\frac{1}{2}} \left(\frac{30}{e}\right)^5 \approx 17,110,222$.

5. (a) $\frac{5 \cdot 4 \cdot 3}{5^3} = .48$; (b) $\frac{(n)_m}{n^m}$.

6. $1!\binom{6}{1} + 2!\binom{6}{2} + 3!\binom{6}{3} + 4!\binom{6}{4} + 5!\binom{6}{5} + 6!\binom{6}{6} = 1956$.

7. Assuming the line has an orientation (that is, it has a distinguished 'front'), there are $n!$ possible lines. Otherwise, there are $\frac{n!}{2}$ lines. Assuming only the relative positions of the people in the circle are relevant, there are $(n-1)!$ circles.

8. (a) $4 \times 26^3 \times 10^3$;
   (b) $4 \times 26 \times 25 \times 24 \times 9 \times 8 \times 7$;
   (c) we interpret the second 'or' to mean 'or both' and get
   $4 \times 5 \times 26^2 \times 10 \times 9 \times 8 + 4 \times 26^3 - 4 \times 5 \times 26^2$
   (we subtract $4 \times 5 \times 26^2$ cases to avoid counting those twice).

Chapter 7

1. (a) For a set of $2n$ people containing n men and n women, choosing a subset of n people can be accomplished by choosing k men and $n - k$ women. There are $\binom{n}{k}\binom{n}{n-k} = \binom{n}{k}^2$ ways of doing this, and of course these must be summed, with k ranging from 0 to n.

   (b) Left-hand side: From a set of n people, choose a team of k people and pick a captain. Add up the total number of possible captained teams (with at least one member). Right-hand side: Pick the captain first, then choose all possible teams from the people remaining. Analytic proof: In the Binomial Theorem (Eq. (7.3)), let $a = t$ and $b = 1$, and differentiate with respect to t.

2. There are $\binom{20}{2}$ ways of choosing the two people, 365 possibilities for the shared birthday, and $(364)_{18}$ ways for the remaining 18 people to have birthdays:

   $$\frac{\binom{20}{2}\cdot 365 \cdot (364)_{18}}{365^{20}} \approx .323 \text{ (using Stirling's formula)}.$$

3. (b) The player may receive the ten clubs in $\binom{13}{10}$ ways and the remaining cards in $\binom{39}{3}$ ways. The probability is $\dfrac{\binom{13}{10}\binom{39}{3}\binom{39}{13\ 13\ 13}}{\binom{52}{13\ 13\ 13\ 13}}$.

   (c) $\dfrac{\binom{4}{3}\binom{48}{10}\binom{39}{13\ 13\ 13}}{\binom{52}{13\ 13\ 13\ 13}} + \dfrac{\binom{4}{4}\binom{48}{9}\binom{39}{13\ 13\ 13}}{\binom{52}{13\ 13\ 13\ 13}}$.

5. (b) Exactly two cells empty:

   $$\frac{\binom{n}{2}\left[\binom{n-2}{1}\binom{n}{3\ 1\ 1\ldots 1} + \binom{n-2}{2}\binom{n}{2\ 2\ 1\ 1\ldots 1}\right]}{n^n}.$$

   (c) Interpreting "a ball" to mean "at least a ball," the probability is $1 - \frac{n}{n^n}$.

7. (a) $\left(\frac{15}{7}\right)^6$.

8. (b) $\binom{n-k}{m-k}$.

   (c) $\sum_{i=l}^{k}\binom{k}{i}\binom{n-k}{m-i}$.

9. (a) $\frac{50}{\binom{100}{2}}$.

   (c) There are $\binom{50}{10}$ ways of choosing the ten states represented and 2^{10} ways of choosing senators from these states. The probability is $\dfrac{2^{10}\binom{50}{10}}{\binom{100}{10}}$.

Chapter 8

1. (a) ($n = 5, \alpha = \frac{1}{3}$.) Denoting the probability distribution by p_a, we have

$$p_a(0) = \frac{32}{243}, \ p_a(1) = \frac{80}{243}, \ p_a(2) = \frac{80}{243}, \ p_a(3) = \frac{40}{243},$$

$$p_a(4) = \frac{10}{243}, \ p_a(5) = \frac{1}{243}.$$

(b) ($n = 5, \alpha = \frac{1}{2}$.) Denoting the probability distribution by p_b, we have

$$p_b(0) = \tfrac{1}{32}, \ p_b(1) = \tfrac{5}{32}, \ p_b(2) = \tfrac{10}{32}, \ p_b(3) = \tfrac{10}{32}, \ p_b(4) = \tfrac{5}{32} \ p_b(5) = \tfrac{1}{32}.$$

2. Suppose that n is even. We have $n = 2k$ for some nonnegative integer k. Define $p_{2k}(x) = \binom{2k}{x}$; we wish to show that $p_{2k}(k)$ is a maximum. We proceed by induction on k. Certainly $p_{2k}(k)$ is a maximum when $k = 0$ or $k = 1$. Suppose the result holds for $k = m$, and consider the case $k = m + 1$. We have

$$p_{2(m+1)}(x + 1) = \binom{2(m + 1)}{x + 1} = \frac{(2m + 2)(2m + 1)}{(2m + 1 - x)(x + 1)} \binom{2m}{x}.$$

By the inductive hypothesis, $\binom{2m}{x}$ is maximized for $m = x$, and a quick calculus argument shows that $\frac{(2m+2)(2m+1)}{(2m+1-x)(x+1)}$ is also maximized for $m = x$. Thus, $p_{2(m+1)}$ has a maximum at $m + 1$. A similar argument can be used to prove that $p(n)$ has two equal maxima at the values $\frac{n-1}{2}$ and $\frac{n+1}{2}$.

3. $\binom{10}{2\ 3\ 5}(\frac{1}{6})^2(\frac{1}{6})^3(\frac{2}{3})^5$ (under some simplifying assumption).

4. Geometric with parameter $\alpha = \frac{5}{6}$.

8. Bernoulli with parameter $\alpha = \frac{1}{5}$.

9. (a) Binomial with parameters $n = 20$, $\alpha = \frac{1}{5}$.

10. (a) Geometric with parameter $\alpha = \frac{1}{5}$.

11. (a) $p_1(n) = \binom{n-1}{r-1}\alpha^r(1 - \alpha)^{n-r}$; (b) use the Binomial Theorem.

12. (a) There are $\binom{M}{i}$ ways of selecting i black balls, $\binom{N-M}{n-i}$ ways of selecting $n - i$ white balls, and $\binom{N}{n}$ total ways of selecting all n balls, so with $0 \le i \le n$ and $i \le M$,

$$p_2(i) = \frac{\binom{M}{i}\binom{N-M}{n-i}}{\binom{N}{n}}.$$

Chapter 9

2. If $\alpha = 1$, then $\mathbb{P}(\{c\}) = 0$ and there is little more to say about $\mathbb{P}(\{a\})$ and $\mathbb{P}(\{b\})$. If $\alpha = 0$, then $\mathbb{P}(\{b\}) = 0$. Otherwise, we have
$$\alpha = \mathbb{P}(\{a,b\} \,|\, \{b,c\}) = \frac{\mathbb{P}(\{a,b\} \cap \{b,c\})}{\mathbb{P}(\{b,c\})} = \frac{\mathbb{P}(\{b\})}{\mathbb{P}(\{b,c\})},$$
so $\mathbb{P}(\{b\}) = \alpha \mathbb{P}(\{b,c\}) = \alpha(\mathbb{P}(\{b\}) + \beta)$, i.e., $\mathbb{P}(\{b\}) = \frac{\alpha\beta}{1-\alpha}$.

   Thus, $\mathbb{P}(\{b,c\}) = \mathbb{P}(\{b\}) + \mathbb{P}(\{c\}) = \frac{\beta}{1-\alpha}$, so $0 < \alpha + \beta < 1$.

3. We take the phrase 'either C or D' to mean one or the other, but not both. (The solution for the other interpretation is left to the reader.) In this case, the answer is $\frac{2}{3}$. How does this compare to the probability (not conditional) that B is released?

4. (a) $\frac{1}{3}$;
   (b) $\frac{2}{3}$.

5. Let A be the event 'dies of Asian flu.' Let B be the event 'has been vaccinated,' so $\bar{B}$ is the event 'hasn't been vaccinated.' Assume that 17% of the population has died of Asian flu. By the Theorem of Total Probabilities, $\mathbb{P}(A) = \mathbb{P}(A \,|\, B)\mathbb{P}(B) + \mathbb{P}(A \,|\, \bar{B})\mathbb{P}(\bar{B})$, which for the information given yields $0.17 = \mathbb{P}(A \,|\, B)(0.20) + \frac{1}{5}(0.80)$. Thus, Archie's chance of survival, given that he undergoes vaccination, is $\frac{1}{20}$.

7. Use induction.

8. This one may be tricky. You might think it is .5. It is not. Proceeding with care, we take as our sample space the set $\Omega = \{(b,b), (b,g), (g,b), (g,g)\}$, where b and g denote 'boy' and 'girl,' respectively. Define the two sets $A = \{\omega \,|\, \text{there is a boy in } \omega\}$ and $B = \{\omega \,|\, \text{there are two boys in } \omega\}$. We have then $B \subset A$, and so $\mathbb{P}(B \,|\, A) = \frac{\mathbb{P}(B \cap A)}{\mathbb{P}(A)} = \frac{\mathbb{P}(B)}{\mathbb{P}(A)} = \frac{1/4}{3/4} = \frac{1}{3}$.

Chapter 10

1. (a) $1 + xy - x - y = (1-x)(1-y)$; (b) $x(1-y)$; (d) $1 - y(1-x)$.

2. $\bar{A}$ and $\bar{B}$ are independent by Theorem 10.4.

3. Use the Theorem of Total Probabilities. The converse holds as long as $\mathbb{P}(A) \neq 0$.

4. (a) not independent; (b) independent; (c) independent.

5. (a) $\Omega = \{(i,j) \,|\, 1 \leq i \leq j \leq 6\}$.

   (b) Denoting the probability distribution on Ω by p, we have
   $p(1,j) = \frac{1}{36}$ for each j, $1 \leq j \leq 6$,
   $p(2,j) = \frac{1}{30}$ for each j, $2 \leq j \leq 6$,
   $p(3,j) = \frac{1}{24}$ for each j, $3 \leq j \leq 6$,

$p(4, j) = \frac{1}{18}$ for each j, $4 \le j \le 6$,
$p(5, j) = \frac{1}{12}$ for $j = 5, 6$,
$p(6, 6) = \frac{1}{6}$.

(c) $\mathbb{P}(\text{Nero wins}) = 5(\frac{1}{36}) + 4(\frac{1}{30}) + 3(\frac{1}{24}) + 2(\frac{1}{18}) + \frac{1}{12}$,
$\mathbb{P}(\text{Nero wins exactly \$2}) = \frac{1}{36} + \frac{1}{30} + \frac{1}{24} + \frac{1}{18}$.

(d) $\mathbb{P}(\text{Archie even} \cap \text{Nero odd}) = \frac{11}{90}$,
$\mathbb{P}(\text{Archie even}) = \frac{1}{2}$,
$\mathbb{P}(\text{Nero odd}) = \frac{402}{1080}$.
Since $\frac{11}{90} \ne (\frac{1}{2})(\frac{402}{1080})$, the events are not independent.

(e) Nero's expected gain on a turn is $\frac{13}{30}$, and Archie's is $-\frac{13}{30}$.

6. $\mathbb{P}(\mathbf{H}_1|D) = \frac{\mathbb{P}(D|\mathbf{H}_1)\mathbb{P}(\mathbf{H}_1)}{\mathbb{P}(D|\mathbf{H}_1)\mathbb{P}(\mathbf{H}_1)+\mathbb{P}(D|\mathbf{H}_2)\mathbb{P}(\mathbf{H}_2)} = \frac{(\frac{1}{6})^3(.7)}{(\frac{1}{6})^3(.7)+(\frac{3}{8})^2(\frac{1}{8})(.3)} \approx .38.$

So $\mathbb{P}(\mathbf{H}_2|D) \approx .62$, and $\mathbf{H}_2$ is favored (despite the very heavy a priori bias in favor of $\mathbf{H}_1$).

7. Let $\Omega = \{a, b, c, d\}$ be the sample space, with each outcome equally likely. Let event $A_1 = \{a, b\}$, event $A_2 = \{a, c\}$, and event $A_3 = \{a, d\}$. Then, A_1, A_2, and A_3 are pairwise independent, since for $i \ne j$, we have $\frac{1}{4} = \mathbb{P}(\{a\}) = \mathbb{P}(A_i \cup A_j) = \mathbb{P}(A_i)(A_j) = (\frac{1}{2})^2$. However, we also have $\frac{1}{4} = \mathbb{P}(\{a\}) = \mathbb{P}(A_1 \cup A_2 \cup A_3) \ne \mathbb{P}(A_1)\mathbb{P}(A_2)\mathbb{P}(A_3) = (\frac{1}{2})^3$, so Eq. (10.11) is not satisfied.

8. $\frac{4}{9}$. Use Bayes' Theorem.

Chapter 11

2. Differentiating the logarithm of (11.4) with respect to a, $0 < a < 1$, one obtains
$$\frac{N_1 + N_2}{a} - \frac{N_2 + 2N_3}{1 - a}.$$

Setting this equal to 0 and solving for a gives the result in (11.4).

(Note that differentiating the logarithm of (11.4) twice with respect to a gives a negative function, so the value obtained in (11.4) is a maximum.)

3. (b) $\mathbb{P}_{a,\theta}(\{(\bar{C}, C, C)\}) = a(1 - \theta)^2$
$\mathbb{P}_{a,\theta}(\{(\bar{C}, \bar{C}, C)\}) = a\theta(1 - \theta) + (1 - a)a(1 - \theta)$
$\mathbb{P}_{a,\theta}(\{(\bar{C}, C, \bar{C})\}) = a(1 - \theta)\theta$
$\mathbb{P}_{a,\theta}(\{(\bar{C}, \bar{C}, \bar{C})\}) = a\theta^2 + (1 - a)a\theta + (1 - a)^2$

(c) For an experiment consisting of N observations, with N_1 of them being $(\bar{C}, \bar{C}, \bar{C})$, N_2 of them $(\bar{C}, \bar{C}, C)$, N_3 of them $(\bar{C}, C, C)$, and $N_4 = N - N_1 - N_2 - N_3$ of them $(\bar{C}, C, \bar{C})$, the likelihood function is

$$[a\theta^2 + (1-a)a\theta + (1-a)^2]^{N_1} [a\theta(1-\theta) + (1-a)a(1-\theta)]^{N_2} [a(1-\theta)^2]^{N_3} [a(1-\theta)\theta]^{N_4}.$$

Taking logarithms, differentiating with respect to a and to θ, and setting the respective derivatives equal to 0 yields a system of two equations in two unknowns a and θ. The solution of this system yields maximum likelihood estimates for a and θ given the data.

Chapter 12

1. (a) Yes; (b) Yes; (c) $\frac{1}{2}$; (d) $\frac{1}{3}$; (e) $\frac{1}{3}$.

2. $\mathbf{X} = \mathbf{X}_1 + \ldots + \mathbf{X}_n$, where $\mathbf{X}_i, 1 \leq i \leq n$ is Bernoulli with parameter α. If the r.v.'s $\mathbf{X}_i$ are independent, then $\mathbf{X}$ is a binomial r.v. with parameters n and α.

4. $\frac{1}{10}$

5. (a) $\sum_{i=0}^{7} \binom{20}{i} \alpha^i (1-\alpha)^{20-i}$

   (b) $\sum_{i=0}^{5} \binom{17}{i} \alpha^i (1-\alpha)^{17-i}$.

   (c) Compare $\mathbb{P}(\sum_{n=1}^{20} \mathbf{Z}_n \leq 7)$ and $\mathbb{P}(\mathbf{Z}_1 + \mathbf{Z}_2 + \mathbf{Z}_3 = 2)$.

   (d) The probability is $\sum_{i=0}^{\infty} (1-\alpha)^{2i} \alpha$, which for $\alpha > 0$ equals $\frac{1}{2-\alpha}$ and is always greater than $\frac{1}{2}$.

6. No.

7. (a) $\alpha^5(1-\alpha)$; (b) $\alpha^{n-k}(1-\alpha)$; (c) α^5; (d) α^{n-k}.

8. The proofs of parts (a), (c), and (d) are similar to the proof given for Theorem 12.13. For part (b), notice that, for $r \in \mathbb{R}$ (and $(\Omega, \mathcal{F}, \mathbb{P})$ the probability space),

$$\{\omega \in \Omega \,|\, \mathbf{X}(\omega) + \mathbf{Y}(\omega) \leq r\}$$
$$= \cup_{y \in \mathbf{Y}(\Omega)} (\{\omega \in \Omega \,|\, \mathbf{X}(\omega) \leq r - y\} \cap \{\omega \in \Omega \,|\, \mathbf{Y}(\omega) = y\}).$$

9. Let $\mathbf{W}$ and $\mathbf{V}$ be discrete random variables in a probability space $(\Omega, \mathcal{F}, \mathbb{P})$. Use the countable additivity of $\mathbb{P}$ (twice) and (12.12) to show that, for any $x, y \in \mathbb{R}$,

$$\mathbb{P}(\mathbf{W} \leq x)\mathbb{P}(\mathbf{V} \leq y) = \sum_{\substack{i \in \mathbf{W}(\Omega) \\ i \leq x}} \sum_{\substack{j \in \mathbf{V}(\Omega) \\ j \leq y}} \mathbb{P}(\mathbf{W} = i)\mathbb{P}(\mathbf{V} = j)$$

$$= \sum_{\substack{i \in \mathbf{W}(\Omega) \\ i \leq x}} \sum_{\substack{j \in \mathbf{V}(\Omega) \\ j \leq y}} \mathbb{P}(\mathbf{W} = i, \mathbf{V} = j) = \mathbb{P}(\mathbf{W} \leq x, \mathbf{V} \leq y).$$

Chapter 13

3. $F_\mathbf{Y}(x) = 0$ if $x < 2$;
   $F_\mathbf{Y}(x) = \frac{1}{36} + \frac{2}{36} + \cdots + \frac{i-1}{36}$ if $i \le x < i+1$, for $2 \le i \le 7$;
   $F_\mathbf{Y}(x) = \frac{1}{36} + \frac{2}{36} + \cdots + \frac{6}{36} + \frac{5}{36}$ if $8 \le x < 9$;
   $F_\mathbf{Y}(x) = \frac{1}{36} + \frac{2}{36} + \cdots + \frac{6}{36} + \frac{5}{36} + \frac{4}{36}$ if $9 \le x < 10$;
   $F_\mathbf{Y}(x) = \frac{1}{36} + \frac{2}{36} + \cdots + \frac{6}{36} + \frac{5}{36} + \frac{4}{36} + \frac{3}{36}$ if $10 \le x < 11$;
   $F_\mathbf{Y}(x) = \frac{1}{36} + \frac{2}{36} + \cdots + \frac{6}{36} + \frac{5}{36} + \frac{4}{36} + \frac{3}{36} + \frac{2}{36}$ if $11 \le x < 12$;
   $F_\mathbf{Y}(x) = 1$ if $x \ge 12$.

4. $F_{\mathbf{Z}_n}(x) = 0$ if $x < 0$;
   $F_{\mathbf{Z}_n}(x) = \sum_{i=0}^{k} \binom{n}{i}(\frac{1}{2})^n$ if $k \le x < k+1$, for $k \in \mathbb{N}$, $k < n$;
   $F_{\mathbf{Z}_n}(x) = 1$ if $x \ge n$.

5. (a) Both are distribution functions. (b) Let $\mathbf{X}$ be a random variable. If G is a continuous, increasing function that maps $F_\mathbf{X}(\mathbb{R})$ (note that $]0,1[\subseteq F_\mathbf{X}(\mathbb{R}) \subseteq [0,1]$) onto itself, then $G(F_\mathbf{X})$ is a distribution function.

   The conditions in this statement may be made weaker.

6. Denoting the range of $\mathbf{V}$ by $\mathbf{V}(\Omega)$, it follows from the Theorem of Total Probabilities that $F_{\mathbf{V}+\mathbf{T}}(x) = \sum_{y \in \mathbf{V}(\Omega)} F_\mathbf{T}(x - y)$ for all $x \in \mathbb{R}$.

7. (a) Examine $\{\omega \in \Omega \,|\, F_\mathbf{Z}(\mathbf{Z}(\omega)) \le y\}$ for any $y \in \mathbb{R}$.
   (b) $F_{F_\mathbf{Z}(\mathbf{Z})}(x) = \mathbb{P}(F_\mathbf{Z}(\mathbf{Z}) \le x) = \mathbb{P}\{\omega \in \Omega \,|\, F_\mathbf{Z}(\mathbf{Z}(\omega)) \le x\}$, which equals

$$
\begin{cases}
0 & \text{if } \frac{1}{2} < x; \\
1 - (\frac{1}{2})^n & \text{if } 1 - (\frac{1}{2})^n \le x < 1 - (\frac{1}{2})^{n+1} \quad \text{for some } n \in \mathbb{N}; \\
1 & \text{if } x \ge 1.
\end{cases}
$$

   (c) See Theorem 14.4 in this connection.

8. The proof is straightforward. Note that $\sum_i p(x_i) = F_\mathbf{X}(x_1) + (F_\mathbf{X}(x_2) - F_\mathbf{X}(x_1)) + (F_\mathbf{X}(x_3) - F_\mathbf{X}(x_2)) + \cdots + (F_\mathbf{X}(x_n) - F_\mathbf{X}(x_{n-1})) = F_\mathbf{X}(x_n) = 1$.

Chapter 14

1. (a) $\frac{2}{3}$;   (b) $1 - \frac{1}{e}$;   (c) 0.84.

2. By the memory-less property of the exponential, this probability is equal to the probability that Archie gets to the track within 6 minutes of leaving the brownstone, that is, $1 - e^{-0.6}$.

3. $\frac{1}{2}$; $\frac{1}{4}$.

4. The value of the distribution function at $0 < t < 1$ is given by $\mathbb{P}(\max\{\mathbf{X}_1, \mathbf{X}_2\} \le t) = \mathbb{P}(\mathbf{X}_1 \le t, \mathbf{X}_2 \le t)$, which is simply $\mathbb{P}(\mathbf{X}_1 \le t)\mathbb{P}(\mathbf{X}_2 \le t)$ by independence (cf. 12.16).

5. Hint: Look at $\mathbb{P}(\min\{\mathbf{Y_1}, \mathbf{Y_2}\} > t)$.

6. The hazard function $h(x) = \frac{f\mathbf{x}}{1 - F\mathbf{x}}$ is constant when $\mathbf{X}$ is exponential. Suppose that $\mathbf{X}$ measures the life time of an item of equipment; then, $h(x)\,dx$ is the probability that the item will fail between x and $x + dx$ given that it is still functioning at time x.

7. (b) $F_e\mathbf{x}(t) = 1 - \frac{1}{t}$ for $t \geq 1$, and $F_e\mathbf{x} = 0$ otherwise.

Chapter 15

1. Let $\mathbf{X}$ be a binomially distributed r.v. with parameters α and n.

   PROOF 1. Let $\mathbf{X}_i$, $1 \leq i \leq n$ be n independent, Bernoulli random variables with parameter α. Then $E(\mathbf{X}_i) = \alpha$, so

   $$E(\mathbf{X}) = E(\sum_{i=1}^{n} \mathbf{X}_i) = \sum_{i=1}^{n} E(\mathbf{X}_i) = n\alpha.$$

   PROOF 2.

   $$E(\mathbf{X}) = \sum_{k=0}^{n} k \binom{n}{k} \alpha^k (1 - \alpha)^{n-k}$$

   $$= n\alpha \sum_{k=1}^{n} \binom{n-1}{k-1} \alpha^{k-1} (1 - \alpha)^{n-k}$$

   $$= n\alpha \sum_{i=0}^{n-1} \binom{n-1}{i} \alpha^i (1 - \alpha)^{n-1-i} \qquad\qquad \text{for } i = k - 1$$

   $$= n\alpha[(\alpha + (1 - \alpha))^{n-1}] \qquad\qquad \text{(by the Binomial Theorem)}$$

   $$= n\alpha.$$

2. The expectation is 1. Write the random variable $\mathbf{N}$ giving the number of letters that end up in the correct envelopes as $\mathbf{N} = \sum_{i=1}^{n} \mathbf{N}_i$, where

   $$\mathbf{N}_i = \begin{cases} 1 & \text{if the } i\text{th letter ends up in the correct envelope} \\ 0 & \text{otherwise} \end{cases}$$

3. The expectation is $\frac{93}{16}$, and the variance is $\frac{263}{256}$.

4. The expectation is $\frac{7}{4}$, and the variance is $\frac{3}{16}$.

5. No.

6. With the sample space $\mathbb{N}$ having the geometric distribution with parameter $\alpha = \frac{1}{2}$, define $\mathbf{X}$ on $\mathbb{N}$ by $\mathbf{X}(n) = (-1)^n \frac{2^n}{n}$. Note that $E(\mathbf{X}) = \sum_{n=1}^{\infty} (-1)^n \frac{1}{n}$, which converges, but $E(\mathbf{X}^2) = \sum_{n=1}^{\infty} \frac{2^n}{n^2}$, which diverges.

7. For computing the expectation and variance of a geometric random variable, note that, for $\alpha \in]0, 1[$,

$$\sum_{n=1}^{\infty} n\alpha(1-\alpha)^{n-1} = \alpha \frac{d}{d\alpha} \sum_{n=1}^{\infty} -(1-\alpha)^n$$

and

$$\sum_{n=1}^{\infty} n^2\alpha(1-\alpha)^{n-1} = \alpha \frac{d}{d\alpha} \sum_{n=1}^{\infty} -n(1-\alpha)^n.$$

For computing the variance of a Poisson random variable, note that, for $\lambda > 0$,

$$\sum_{k=0}^{\infty} k^2 e^{-\lambda} \frac{\lambda^k}{k!} = \lambda e^{-\lambda} \frac{d}{d\lambda} \left[\sum_{k=1}^{\infty} \frac{\lambda^k}{(k-1)!} \right].$$

8. Yes. Let $\mathbf{X}$ be a r.v. such that $\mathbb{P}(\mathbf{X} = x) = \frac{1}{3}$ for $x = 1, 2, -3$, and $\mathbb{P}(\mathbf{X} = x) = 0$ otherwise. Then $E(\mathbf{X}) = 0 = E(-\mathbf{X})$ and $Var(\mathbf{X}) = Var(-\mathbf{X})$, but $As(\mathbf{X}) = -As(-\mathbf{X})$.

9. $m_e = -\frac{1}{3} \ln \frac{1}{2}$.

10. No. Can you explain why?

Chapter 16

4. $E(\mathbf{U}) = 2$, $E(\mathbf{V}) = 4$, $Var(\mathbf{U}) = \frac{2}{3}$, $Var(\mathbf{V}) = \frac{4}{3}$, and $Cov(\mathbf{U}, \mathbf{V}) = \frac{24}{9}$.

5. Note that $\mathbf{M}_i = \sum_{\ell=1}^{n} \mathbf{X}_{i,\ell}$, where

$$\mathbf{X}_{i,\ell} = \begin{cases} 1 & \text{if type } i \text{ is drawn on trial } \ell \\ 0 & \text{otherwise,} \end{cases}$$

and that $\mathbf{M}_i$ is binomial with parameters n and α_i. It follows that $E(\mathbf{M}_i) = n\alpha_i$ and that $Var(\mathbf{M}_i) = n\alpha_i(1-\alpha_i)$. Show that $Cov(\mathbf{M}_i, \mathbf{M}_j) = -n\alpha_i\alpha_j$ for $i \neq j$. (Why would the covariance be negative?)

6. For part (b), note that $Var(\sum_{i=1}^{n} \mathbf{X}_i) = Cov(\sum_{i=1}^{n} \mathbf{X}_i, \sum_{j=1}^{n} \mathbf{X}_j)$.

Chapter 17

1. From Theorem 17.1 and Example 15.3, we have that $\mathbb{P}(e^{\mathbf{X}} \geq 2\frac{\lambda}{\lambda-1}) \leq \frac{1}{2}$, that is, $\mathbb{P}(\mathbf{X} \geq \ln 4) \leq \frac{1}{2}$. The exact computation gives $\mathbb{P}(\mathbf{X} \geq \ln 4) \leq \frac{1}{16}$.

3. Yes. We still have Eq. (17.7) holding if the random variables are uncorrelated; see Problem 6(b), Chapter 16.

5. (a) $\frac{2}{3}$; (b) $1 - \Phi(1.58) \approx .06$.

7. $\frac{1}{n^2}$.

8. Approximately .037.

9. Yes. Use Theorem 17.1.

*10. PROOF. With $\bar{A} = \Omega \setminus A$, observe that

$$\int_A g(x) \, d\mathbb{P}(x \le \mathbf{X}) \le \max g\left(\mathbf{X}(A)\right) \mathbb{P}(A),$$

$$\int_{\bar{A}} g(x) \, d\mathbb{P}(x \le \mathbf{X}) \le g(a).$$

Adding these two equations gives

$$E\left(g(\mathbf{X})\right) \le \max g\left(\mathbf{X}(A)\right) \mathbb{P}(A) + g(a),$$

which is equivalent to the result of the problem because $\max g\left(\mathbf{X}(A)\right) > 0$.

Chapter 18

1. $M_{a\mathbf{X}+b}(\theta) = E[e^{\theta(a\mathbf{X}+b)}] = e^{\theta b} E[e^{(\theta a)\mathbf{X}}] = e^{\theta b} M_{\mathbf{X}}(\theta a)$.

4. For $\mathbf{X} = \sum_{i=1}^{n} \mathbf{X}_i$ (a gamma r.v. with parameters n and λ) and for $\mathbf{Y} = \sum_{j=1}^{m} \mathbf{Y}_j$ (a gamma r.v. with parameters m and λ), with $\mathbf{X}$ and $\mathbf{Y}$ independent, we have by Theorem 18.6
$M_{\mathbf{X}+\mathbf{Y}}(\theta) = M_{\mathbf{X}}(\theta) M_{\mathbf{Y}}(\theta) = M_{\mathbf{X}_1}(\theta) \cdots M_{\mathbf{X}_n}(\theta) M_{\mathbf{Y}_1}(\theta) \cdots M_{\mathbf{Y}_m}(\theta)$.

5. (a) Use the facts that $a\mathbf{X}$ and $b\mathbf{Y}$ are normal (Theorem 14.10) and independent (Theorems 12.14 and 12.15). (b) Show that $Cov(a\mathbf{X}+b\mathbf{Y}, a\mathbf{X} - b\mathbf{Y}) = 0$. (Use the result in Problem 6(a), Chapter 16.)

6. (b) No. In the case that $m = n$, for instance, it can be shown that $(\alpha_{\mathbf{X}} e^{\theta} + 1 - \alpha_{\mathbf{X}})^m (\alpha_{\mathbf{Y}} e^{\theta} + 1 - \alpha_{\mathbf{Y}})^m = (\gamma e^{\theta} + 1 - \gamma)^k$ for some γ and k implies that $\alpha_{\mathbf{X}} = \alpha_{\mathbf{Y}}$.

9. Independence and the M.G.F. for a Poisson random variable give $M_{\mathbf{S}_n}(\theta) = e^{(\lambda + \lambda^2 + \cdots + \lambda^n)(e^{\theta} - 1)}$, and this tends to $e^{\frac{\lambda}{1-\lambda}(e^{\theta} - 1)}$ as $n \to \infty$.

10. No. Examine the product of the M.G.F.'s of independent uniform r.v.'s.

Chapter 19

1. For any $x \in \mathbb{R}$ and $i \in \{1, \ldots, k\}$, note that the set
$\{\omega \in \Omega | \mathbf{X}_i(\omega) \le x\} = \{(m_1, \ldots, m_k) | m_i \le x, \sum_{j=1}^{k} m_j = 1\} \subseteq \Omega$
is a closed subset of $\mathbb{R}^k$, so it is a Borel set.

3. Let $i = 1$ for simplicity. With all sums below being taken over all $(k-1)$-tuples $(m_2, \ldots, m_k)$ such that $\sum_{j \ne 1} m_j = n - m_1$, we have
$\mathbb{P}(\mathbf{X}_1 = m_1) = \sum \mathbb{P}(\mathbf{X}_1 = m_1, \mathbf{X}_2 = m_2, \ldots, \mathbf{X}_k = m_k)$
$= \sum \binom{n}{m_1 \cdots m_k} \alpha_1^{m_1} \cdots \alpha_k^{m_k} = \frac{n!}{m_1!(n-m_1)!} \alpha_1^{m_1} \sum \binom{n-m_1}{m_2 \cdots m_k} \alpha_2^{m_2} \cdots \alpha_k^{m_k}$
$= \frac{n!}{m_1!(n-m_1)!} \alpha_1^{m_1} (1 - \alpha_1)^{n-m}$, where the last equality follows from the Multinomial Theorem (7.9).

5. **X** and **Y** are independent when $\alpha = \frac{1}{2}$ and $\beta_1 = \beta_2$.

6. $Var(\mathbf{X}) = \alpha Var(\mathbf{Y}) + (1 - \alpha)Var(\mathbf{Z}) + \alpha(1 - \alpha)(E(\mathbf{Y}) - E(\mathbf{Z}))^2$.

8. Let $F_{\mathcal{X}}$ be as given in (19.12), with $\mathcal{X} = (\mathbf{X}_i)_{1 \le i \le n}$. Then the marginal distribution of $\mathbf{X}_i$ is given by $F_{\mathbf{X}_i}(x_i) = \mathbb{P}(\mathbf{X}_i \le x_i) = \lim_{\substack{x_j \to \infty \\ j \ne i}} F_{\mathcal{X}}(x_1, \ldots, x_n)$.

10. From Lemma 14.8, it can be shown that $f(x + y) = g(x)h(y)$, etc.

11. Let $D = \alpha(1 - \beta_1)\beta_1^{16} + (1 - \alpha)(1 - \beta_2)\beta_2^{16}$. Then $E(\mathbf{X}|\mathbf{Y} = 17) = D^{-1}[\alpha(1-\beta_1)\beta_1^{16} + 2(1-\alpha)(1-\beta_2)\beta_2^{16}]$ and $Var(\mathbf{X}|\mathbf{Y} = 17) = D^{-1}[\alpha(1-\beta_1)\beta_1^{16} + 4(1-\alpha)(1-\beta_2)\beta_2^{16}] - D^{-2}[\alpha(1-\beta_1)\beta_1^{16} + 2(1-\alpha)(1-\beta_2)\beta_2^{16}]^2$.

15. Beginning with the r.h.s., we have $E\left(\mathbf{Z}^2 - E(\mathbf{Z})^2\right) + (E(\mathbf{Z}) - \alpha)^2 = E(\mathbf{Z}^2) - E(\mathbf{Z})^2 + E(\mathbf{Z})^2 - 2\alpha E(\mathbf{Z}) + \alpha^2 = E\left(\mathbf{Z}^2 - 2\alpha\mathbf{Z} + \alpha^2\right) = E\left((\mathbf{Z} - \alpha)^2\right)$.

Chapter 20

1. Show that the joint normal density $f_{\mathcal{X}}(x_1, x_2)$ can be written as the product of $f_1(x_1)$ and a normal density with parameters $\mu_2 + \rho\frac{\sigma_2}{\sigma_1}(x_1 - \mu_1)$ and $\sigma_2\sqrt{1 - \rho^2}$. Use this and the M.G.F. for a normal random variable (see Eq. (18.3)) to derive (20.3).

2. Setting the r.h.s. of (20.1) equal to the r.h.s. of (20.4) and then successively taking the logarithm of both sides, differentiating with respect to x_1, and differentiating with respect to x_2, we obtain $\frac{\rho}{\sigma_1\sigma_2(1-\rho^2)} = 0$, which holds only if $\rho = 0$.

3. $\frac{\partial^2 M_{\mathcal{X}}}{\partial\theta_1\partial\theta_2} = (\mu_2 + \theta_2\sigma_2^2 + \theta_1\sigma_1\sigma_2\rho)(\mu_1 + \theta_1\sigma_1^2 + \theta_2\sigma_1\sigma_2\rho)M_{\mathcal{X}}(\theta_1, \theta_2) + \sigma_1\sigma_2\rho M_{\mathcal{X}}(\theta_1, \theta_2)$. Thus, because $M_{\mathcal{X}}(0, 0) = 1$, we have

$$E(\mathbf{X}_1\mathbf{X}_2) = \frac{\partial^2 M_{\mathcal{X}}}{\partial\theta_1\partial\theta_2}\Big|_{\theta_1=\theta_2=0} = \mu_1\mu_2 + \sigma_1\sigma_2\rho.$$

5. (a) .023; (b) .092; (c) .68.

6. No. Consider $f_{\mathcal{X}}(x, y) = \frac{1}{2\pi}e^{-\frac{1}{2}(x^2+y^2)}[1 + xy\, e^{-\frac{1}{2}(x^2+y^2-2)}]$. This counterexample is proposed by Hogg and Craig (1995, p. 151). Check that this equation defines a bivariate density, with normal marginals.

8. $1 - \Phi\left(\frac{-(\mu_1-\mu_2)}{\sigma_1^2+\sigma_2^2-2\rho\sigma_1\sigma_2}\right)$, where Φ is the distribution function for a standard normal r.v.

Chapter 21

1. The probability that the subject responds correctly on trial 3, given that she responded correctly on trials 1 and 2, is

$$\frac{\gamma^3(1-\theta)^2 + \gamma^2\theta(1-\theta) + \gamma\theta}{\gamma^2(1-\theta) + \gamma\theta},$$

whereas the probability that she responds correctly on trial 3, given that she responded correctly on trial 2, is

$$\frac{\gamma^3(1-\theta)^2 + \gamma^2\theta(1-\theta) + \gamma\theta + \gamma^2(1-\gamma)(1-\theta) + \gamma(1-\gamma)\theta}{\gamma^2(1-\theta) + \gamma\theta + \gamma(1-\gamma)}.$$

These two probabilities do not coincide for any values of the parameters γ and θ satisfying $0 < \gamma < 1$ and $0 < \theta < 1$.

3. The probability that the subject is in state **M** on trial n, given the states on previous trials $n_1 < n_2 < \ldots < n_j < n$, is 1 if the subject is in state **M** on trial n_j and $1 - [1 - \gamma\eta_{n_j}][1 - \gamma\eta_{n_j+1}]\cdots[1 - \gamma\eta_{n-1}]$ otherwise. The (one-step) transition matrix is

Trial $n + 1$

		M	**N**
Trial n	**M**	1	0
	N	$\gamma\eta_n$	$1 - \gamma\eta_n$

6. The probability that she is in state **N** on trial 10 is $(1 - \gamma\theta)^3$, and the probability that she is in state **M** on trial 10 is $\gamma\theta + (1-\gamma\theta)\gamma\theta + \gamma\theta(1-\gamma\theta)^2$.

7. (a) $\begin{pmatrix} p & 1-p \\ 1-p & p \end{pmatrix}$

8. (a) $\begin{pmatrix} 0 & 1 & 0 \\ \frac{1}{4} & \frac{1}{2} & \frac{1}{4} \\ 0 & 1 & 0 \end{pmatrix}$; (b) $\frac{3}{4}$; (c) $\frac{15}{64}$.

9. The right side of (21.8) is equal to $\frac{3\gamma(1-\gamma)}{1+\gamma}$.

Chapter 22

1. Let J be an arbitrary index set, and let $\cap_{a \in J} B_a$ be an intersection of closed sets of states. Suppose $k \in \cap_{a \in J} B_a$. If there is a state j such that $k \rightarrowtail j$, then $j \in \cap B_a$; indeed if $j \notin B_c$ for some $c \in J$, then B_c is not closed, a contradiction. The closure of $\{2, 5\}$ is $\{2, 4, 5, 6\}$. The closure of $\{4, 5, 6\}$ is $\{4, 5, 6\}$.

2. Denote the closure of B by $cl(B)$. Since $cl(B)$ is an intersection of closed sets, it is closed. Thus, $cl(B)$ contains (at least) all states accessible for some state in B. If $cl(B)$ were to contain a state j that was not accessible from any other state $i \in B$, then $cl(B) \setminus \{j\}$ would be a closed set strictly contained by $cl(B)$, which contradicts the minimality of $cl(B)$.

7. (a) $\mathbb{S} = \{\{1,2,3\}, \{4\}, \{5,6\}\}$. (b) $\{1,2,3\} \precsim \{4\} \precsim \{5,6\}$. Thus, $\{5,6\}$ is ergodic and the other two classes are transient. There are no absorbing states. (c) $f(1) = f(2) = f(3) = \frac{1}{6}$; $f(4) = \frac{1}{2}$; $f(5) = f(6) = 1$.

8. If i and j are mutually accessible, with i persistent, there exist n and m such that $\tau_m(i,j) > 0$ and $\tau_n(j,i) > 0$. Applying (21.10) twice, we get

$$\tau_{n+k+m}(j,j) \geq \tau_{n+k}(j,i)\tau_m(i,j) \geq \tau_n(j,i)\tau_k(i,i)\tau_m(i,j), \qquad \text{and so}$$

$$\sum_{k=1}^{\infty} \tau_{n+k+m}(j,j) \geq \sum_{k=1}^{\infty} \tau_n(j,i)\tau_k(i,i)\tau_m(i,j) = \tau_n(j,i)\tau_m(i,j)\sum_{k=0}^{\infty}\tau_k(i,i) = \infty.$$

9. An ergodic class is closed and so is equal to its closure. A transient class is not closed, so it is a proper subset of its closure.

Chapter 23

1. $\xi(i,j) = \begin{cases} 1-\gamma & \text{if } j = i+1 \text{ and } -98 \leq j \leq 100, \\ \gamma & \text{if } j = i-1 \text{ and } -100 \leq j \leq 98, \\ 0 & \text{in all other cases.} \end{cases}$

2. Yes. All paths have probability $\theta^5 \gamma^4$.

3. (a) $\bigcap_{n=0}^{\infty} \{w \in \Omega \mid q < \mathbf{X}_n(w) < r\}$.
   (c) $\bigcup_{i=0}^{\infty}\left(\bigcap_{n=0}^{i}\{w \in \Omega \mid \mathbf{X}_n(w) = 0\} \cap (\bigcap_{n=i+1}^{\infty}\{w \in \Omega \mid \mathbf{X}_n(w) > 0\})\right) \cup \bigcup_{i=0}^{\infty}\left(\bigcap_{n=0}^{i}\{w \in \Omega \mid \mathbf{X}_n(w) = 0\} \cap (\bigcap_{n=i+1}^{\infty}\{w \in \Omega \mid \mathbf{X}_n(w) < 0\})\right)$.

5. Note that $\lim_{\frac{\gamma}{\theta} \to 1} \frac{r(\gamma/\theta)^{r-1} - k(\gamma/\theta)^{k-1}}{r(\gamma/\theta)^{r-1}} = 1 - \frac{k}{r}$.

7. (a) If Peter starts with k dollars and the probability of a head is γ, then the probability that Peter loses all of his money before he wins r dollars $(r \geq k)$ is $\begin{cases} \frac{[\gamma/(1-\gamma)]^r - [\gamma/(1-\gamma)]^k}{[\gamma/(1-\gamma)]^r - 1} & \text{if } \gamma \neq \frac{1}{2}, \\ 1 - \frac{k}{r} & \text{if } \gamma = \frac{1}{2}. \end{cases}$
   (b) .999; (c) .969; (d) .500.

8. (b) 43.13; (c) 27.

10. If n is odd, then $\mathbb{P}(\mathbf{X}_n = 0) = 0$. Otherwise, there are $\binom{n}{n/2}$ paths which return to the origin in n steps, so $\mathbb{P}(\mathbf{X}_n = 0) = \binom{n}{n/2}\left(\frac{1}{2}\right)^n$.

Chapter 24

2. For any real number x, we have from the definition of the exponential function that $e^x = \sum_{n=0}^{\infty} \frac{x^n}{n!}$; hence, $\sum_{k=j}^{\infty} \frac{(\delta\lambda(1-p))^{k-j}}{(k-j)!} = e^{(1-p)\delta\lambda}$, and (24.5) follows from (24.4).

6. Expectation is $\frac{k}{\lambda}$; variance is $\frac{k}{\lambda^2}$.

7. (a) $e^{-60\lambda}$; (b) $e^{-60\lambda p}$; (c) $e^{-60\lambda p}$; (d) λps (e) λps.

8. (a) 10 minutes; (b) $\frac{1}{16}$; (c) $\frac{7}{16}$; (d) 11 minutes.

9. $1 - e^{-\frac{5}{7}}$.

Glossary of Symbols

$\Leftrightarrow$, iff	the logical equivalence; both stand for 'if and only if'
&	the logical conjunction 'and'
$\in$	set membership
$\emptyset$	empty set
$\subseteq$	set inclusion
$\subset$	proper (or strict) set inclusion
$\cup, \cap$	set union and set intersection
$+, \sum$	may stand for the ordinary addition or for the union of disjoint sets
$\setminus$	difference of sets
$f(B)$	if B is a set and f a function, the image of B by f
$|X|$	number of elements (or cardinal number, cardinality) of a set X
2^X	power set of the set X (i.e., the set of all subsets of X)
$X \times Y$	Cartesian product of the sets X and Y
$\mathbb{N}$	the set of all natural numbers $1, \ldots, n, \ldots$
$\mathbb{N}_0$	the set of nonnegative integers $0, 1, \ldots, n, \ldots$
$\mathbb{Q}$	the set of all rational numbers
$\mathbb{R}$	the set of all real numbers
$\mathbb{R}_+$	the set of all nonnegative real numbers, i.e., $\mathbb{R}_+ = [0, \infty[$.
$\mathbb{P}$	a probability measure
Ω	a sample space, not to be confused with:
$\mathcal{S}$	the state space (of a Markov chain)
p	often, a probability distribution
$]x, y[$	open interval of real numbers $\{z \in \mathbb{R} \mid x < z < y\}$
$[x, y]$	closed interval of real numbers $\{z \in \mathbb{R} \mid x \le z \le y\}$
$]x, y], [x, y[$	real, half open intervals
$\mathcal{B}$	the collection of all Borel sets of $\mathbb{R}$ (or of $\mathbb{R}^n$)
$\square$	marks the end of a proof
l.h.s.	abbreviation for 'left-hand side' (of a formula)
r.h.s.	abbreviation for 'right-hand side' (of a formula)
r.v.	abbreviation sometimes used for 'random variable'
$F_{\mathbf{X}}$	distribution function of the random variable $\mathbf{X}$
$f_{\mathbf{X}}$	density function of the random variable $\mathbf{X}$
Φ	distribution function of a standard normal random variable.

Bibliography

J. Aczél. *Lectures on Functional Equations and their Applications*. Academic Press, New York, 1966.

R.C. Atkinson, G.H. Bower, and E.J. Crothers. *An Introduction to Mathematical Learning Theory*. John Wiley & Sons, New York, 1965.

A.T. Barucha-Reid. *Elementary Probability with Stochastic Processes*. Springer, New York, 1974.

P. Beckmann. *Elements of Applied Probability Theory*. Harcourt, Brace and World, New York, 1967.

J. Bernoulli. *Ars Conjectandi*. Publisher Unknown, 1713.

K.L. Chung. *Elementary Probability with Stochastic Processes*. Springer-Verlang, New York, 1974.

D.R. Cox. The analysis of non-Markovian stochastic srocesses by the inclusion of supplementary variables. *Proceedings of the Cambridge Philosophical Society*, 51:1433–1441, 1955.

J.-Cl. Falmagne. Stochastic token theory. *Journal of Mathematical Psychology*, 41(2):129–143, 1997.

J.-Cl. Falmagne and J.-P. Doignon. Stochastic evolution of rationality. *Theory and Decision*, 43:107–138, 1997.

W. Feller. *An Introduction to Probability and its Applications*. Macmillan, New York, 1957.

E.G. Gumbel. Several early bivariate exponential distributions. *Journal of the American Statistical Association*, 55:698–707, 1960.

R.V. Hogg and A.T. Craig. *Introduction to Mathematical Statistics*. 5th ed., Wiley & Sons, New York, 1995.

N.L. Johnson and S. Kotz. *Distributions in Statistics: Continuous Multivariate Distributions*. Wiley & Sons, New York, 1972.

N.L. Johnson, S. Kotz, and N. Balakrishnan. *Continuous Univariate Distributions*. 2nd ed., Wiley & Sons, New York, 1994.

J.G. Kemeny and J.L. Snell. *Finite Markov Chains*. Van Nostrand, Princeton, N.J., 1960.

A.N. Kolmogorov. *Foundations of the Theory of Probability*. Chelsey, New York, 1950.

S. Kotz, N.L. Johnson, and C.B. Read, editors. *Encyclopedia of Statistical Sciences.* 2nd ed, Wiley-Interscience, New York, 1985.

M. Loève. *Probability Theory.* Van Nostrand, Princeton, N.J., 1963.

A.W. Marshall and I. Olkin. A bivariate exponential distribution for the fatal shock event. *Journal of the American Statistical Association,* 62: 30–44, 1967.

B.J. McNeil, S.J. Sox, H.C., and A. Tversky. On the elicitation of preferences for alternative therapies. *New England Journal of Medicine,* 306: 1259–1252, 1982.

M.F. Norman. *Markov Processes and Learning Models.* Academic Press, New York, 1972.

E. Parzen. *Modern Probability and Its Applications.* Holden-Day, San Francisco, 1994a.

E. Parzen. *Stochastic Processes.* Holden-Day, San Francisco, 1994b.

M. Regenwetter, J.-Cl. Falmagne, and B. Grofman. A stochastic model of preferences change and its application to 1992 presidential election panel data. *Psychological Review,* 106(2):362–384, 1999.

H. Robbins. A remark on Stirling's formula. *American Mathematical Monthly,* 62:26–29, 1955.

F.R. Roberts. Measurement theory. In Gian-Carlo Rota, editor, *Encyclopedia of Mathematics and its Applications,* volume 7. Addison-Wesley, Reading, Ma, 1979.

H.L. Royden. *Real Analysis.* MacMillan, New York, 1966.

W. Rudin. *Principles of Mathematical Analysis.* McGraw-Hill, New York, 1964.

J. Stirling. *Methodus Differentialis.* 1730.

W.T. Trotter. *Combinatorics and Partially Ordered Sets: Dimension Theory.* The Johns Hopkins University Press, Baltimore, Maryland, 1992.

A. Tversky and D. Kahneman. The framing of decisions and the psychology of choice. *Science,* 211:453–458, 1981.

A. Tversky and D. Kahneman. Extensional versus intuitive reasoning: the conjunction fallacy in probability judgment. *Psychological Review,* 293–315, 1983.

I. M. Yaglom. *Felix Klein and Sophus Lie—Evolution of the Idea of Symmetry in the Nineteenth Century.* Birkhäuser, Boston - Basel, 1988.

Index